Competitiveness Matters

Competitiveness Matters

Industry and Economic Performance in the U.S.

Candace Howes and Ajit Singh, Editors

Ann Arbor

THE UNIVERSITY OF MICHIGAN PRESS

Published in the United States of America by
The University of Michigan Press
Manufactured in the United States of America
♾ Printed on acid-free paper

2003 2002 2001 2000 4 3 2 1

*A CIP catalog record for this book is available
from the British Library.*

Library of Congress Cataloging-in-Publication Data

Competitiveness matters : industry and economic performance in the U.S. / Candace Howes and Ajit Singh, editors.
p. cm.
Includes bibliographical references.
ISBN 0-472-10983-9 (cloth : alk. paper)
1. United States—Commercial policy. 2. United States—Economic policy—1993–. 3. Industrial policy—United States. 4. Manufacturing industries—Government policy—United States. 5. Technological innovations—Economic aspects—United States. 6. Balance of trade—United States. 7. Competition, International. I. Howes, Candace. II. Singh, Ajit.

HF1455 .C726 1999
338.973—dc21 99-052803

For David M. Gordon and Bennett Harrison

Contents

CHAPTER 1

Introduction

Competitiveness Matters

Candace Howes and Ajit Singh

Krugman: Competitiveness Does Not Matter

In a series of widely read articles and books published over the last several years, Krugman (1994, 1996) decries what he regards as a "dangerous obsession" with international competitiveness, a trend he refers to as "pop internationalism." With pop internationalism he associates the idea that the recent ills of the U.S. economy—eroding real wages, stagnating living standards, rising inequality and unemployment—are the consequence of a major erosion of the industrial base due to international competition. Pop internationalists, according to Krugman, reason that the economic ills of the United States will be remedied only when the United States has regained a productivity edge over its international rivals. But it is folly, Krugman claims, to think that the notion of competitiveness is in any way meaningful when applied to a nation.

Krugman is of course correct to argue that the competitiveness of a nation is conceptually different from that of a corporation. Certainly, if a corporation is uncompetitive, its market position is likely to be unsustainable, forcing it into bankruptcy. There is no similar analogue for a country. Even if the balance of payments position is unsustainable, even in the event of financial collapse, countries—unlike the banks and corporations that may hold their debt—do not cease to exist or go into bankruptcy.

Nor would we challenge Krugman's argument that trade deficits and surpluses are an inappropriate measure of the competitiveness of a country. While a trade deficit may follow from the weak performance of a country's tradable goods sector, it may also be the consequence of a large inflow of foreign investment that is equated with competitive strength. A trade surplus, too, sends an ambiguous signal. It may be due to a low level of national economic activity or to strong export performance.

However, some scholars, understanding full well that the trade balance is not a good measure of competitiveness, have offered an alternative formulation of national competitiveness that is more difficult for Krugman to challenge. Tyson (1992) defines competitiveness as the "ability to produce goods and services that meet the test of international competition, while our citizens enjoy a standard of living that is both rising and sustainable." Tyson's formulation implies that a competitive country is one that is able to produce tradable goods that are in sufficient demand both domestically and in international markets such that trade will be in balance without the country's having to resort to continual depreciation of its currency or to operating at a level of activity below the full potential of the economy.

To this definition of competitiveness, Krugman offers two objections. The first is that for a relatively closed economy such as the United States was in the 1950s, trade is so small relative to GNP that even if a steady currency depreciation is required to balance trade, the effect on the purchasing capacity of the country's citizens would be negligible. Under the conditions of a relatively closed economy, the standard of living is almost entirely a function of the rate of growth of *domestic* productivity, not productivity growth *relative* to competitors.

He admits that if trade were a large part of GNP any currency devaluation to maintain balanced trade could, in principle, have a depressive effect on the rate of growth of real incomes. However, in actual fact, he argues, since U.S. exports are only 10 percent of GNP, relative price adjustments through the exchange rate cannot have a significant effect on overall purchasing power.

Krugman suggests that his critique of the concept of national competitiveness is empirical and is applicable to advanced economies. He writes: "While competitiveness problems could arise in principle, as a practical, empirical matter, the major nations of the world are not to any significant degree in economic competition with each other" (1994, 35). Krugman does not discuss specifically whether or not the concept of national competitiveness is useful in relation to developing countries. However, a clear implication of his analysis is that the concept is likely to be much more applicable to such countries as they are typically more open and less diversified and therefore more subject to terms of trade shocks.

Thus in Krugman's view, the economic problems of industrial countries—unemployment, deindustrialization, slow growth rates of per capita incomes—cannot be attributed to an uncompetitive position in the rivalry between countries (contrary to what the pop internationalists suggest). Weak performance is due to problems internal to the economies—slow productivity growth, the natural tendency in advanced industrial economies for employment to grow faster in services than manufacturing, and problems associated with regulation, social welfare, and monetary restraint.

To these analytical and empirical objections to the concept of competitiveness, Krugman adds important normative strictures. He regards the attention paid to international competitiveness by policymakers and international organizations as a dangerous obsession. This is because he believes that policymakers wrongly tend to view economic interactions between countries as a zero-sum rather than a positive-sum. Such essentially mercantilist misconceptions about the role of trade can, in his view, lead ultimately to protectionism or worse policies.

In his 1994 contribution to the debate, Krugman accepts that strategic trade theory weakens the analytical case for free trade. However, he suggests that even in a world of imperfect competition and increasing returns to scale, the gains a country may reap from strategic restraints on trade are empirically very small. These gains in his view are far outweighed by the dangers the country runs of retaliation, protectionism, and ultimately trade wars. Hence, he suggests abandonment of the competitive metaphor and the rhetoric of competitiveness by world leaders, policymakers, and some misguided economists.[1]

The Limitations of the Krugman Thesis

Underlying Krugman's critique of the concept of national competitiveness is a standard neoclassical model in which the effects of trade on a country's standard of living manifest themselves mainly through changes in the terms of trade brought about by the equilibrating adjustments of exchange rates.[2] Because complete wage-price flexibility is assumed, and because demand for traded goods is assumed to be perfectly elastic at world prices, balance of payments disequilibria, including those that may arise between countries due to differences in rates of productivity growth, can be smoothly resolved by exchange-rate adjustments. Under neoclassical assumptions then, differences in relative productivity growth and the trade imbalances that may follow cannot have any effect on demand, output, employment, or inflation.

However, in the real world of incomplete wage-price flexibility, the adjustment process may be far from smooth. It may entail leapfrogging inflation and considerable adjustment in quantities, that is, in real output and employment. These difficulties may be illustrated by considering the experience of an advanced country (the United Kingdom in the mid-1970s). Following the first oil price shock in 1973–74, the U.K. economy, which was not then a major producer and exporter of oil, suffered an adverse movement in its terms of trade due to the OPEC oil price increase. The size of the shock was estimated to have amounted to about 4 percent of GDP. Instead of a smooth adjustment of the economy through movements in the exchange rates, there was a protracted process that involved redistributive struggles between various social groups over the diminished national pie. The net result was a doubling of the rate of unemployment, a quadrupling of the rate of inflation, a full-blown financial crisis, and ultimately the humiliation for an advanced industrial country of being forced to accept an IMF rescue package, before internal and external equilibria could be restored.[3] Thus even a relatively small terms-of-trade shock can have serious repercussions even in an advanced country for an economy, depending on the dynamics of the adjustment process. The validity of Krugman's analysis of national competitiveness requires an abstraction from such labor market dynamics.

There is a further more serious problem with the Krugman model of equilibrating adjustment between countries through prices, that is, changes in exchange rates. This arises not so much from the abstractions made with respect to the labor market dynamics but, importantly, it is caused by the neglect of certain essential features of the contemporary product markets. In a wide range of manufactured products these markets are characterized by oligopolistic structures. This leads to a situation that competition now takes place to a considerable extent on the basis of nonprice factors such as quality, marketing, design, reliability, and service.

This aspect of international trade is related to the empirical paradox originally observed by Kaldor (1978). He found that for countries like Germany and Japan that increased their share of world markets in manufactures in the 1960s and the early 1970s, prices and costs relative to other countries (expressed in a common currency) rose rather than declined. On the other hand, the share of the United Kingdom and the United States in the world exports of manufactured goods fell despite the fact that their prices and costs relative to other countries were decreasing. Fagerberg (1996, table 1) has updated Kaldor's original analysis for the years 1963 through 1975 to the period 1978 through 1994; he has also extended it to include twelve leading countries for manufactured exports. He finds a positive relationship still exists between world market share and relative unit labor costs.[4] It is notable that the East Asian NICs, which have gained large increases in market share, have recorded rising relative unit labor costs. From a neoclassical perspective, there would appear then to be a generally perverse relationship between a country's world market share and its relative prices. On the other hand, Fagerberg also observed, seemingly consistent with neoclassical theory, a positive correlation between a country's rate of productivity growth and its change in world market share. An important explanation for these observed phenomena lies partly in the fact mentioned earlier, that is, the increasing role of technology and other nonprice factors in international trade. The reason for the positive association between productivity growth and market share is that countries with high rates of productivity growth also have high rates of investment and output growth. Such countries thereby achieve faster turnover of machines, faster technical progress (to the extent that technical progress is embodied in new capital goods), greater learning by doing, and quicker development of new products.[5] As a consequence, as Kaldor (1981, 603) observed: "Basically in a growing world economy the growth of exports is mainly to be explained by the income elasticity of demand of foreign countries for a country's products; but it is a matter of the innovative ability and adaptive capacity of its manufacturers whether this income elasticity will tend to be relatively large or small."

Kaldor's emphasis on technological development as the key nonprice factor in international competition also finds confirmation in the data on R&D analyzed by Fagerberg. He shows there is a significant positive relationship between the change in a country's R&D as a share of GDP and growth in its world market share.[6]

Empirically, then, market share growth is better explained by relative productivity growth (and the associated growth of investment) than by falling relative unit labor costs. These analytical and empirical findings, when coupled with the concept of cumulative causation, have serious implications for the Krugman analysis. It will be recalled that one of Krugman's main points is to suggest that a nation's standard of living is determined by its own long-term productivity growth rather than its productivity growth relative to others (subject to a terms-of-trade effect discussed earlier). However, if countries' relative productivity growth is an indicator of their relative nonprice competitiveness, it means that a country with relatively slow productivity growth will not only have a smaller growth of market share but that because of cumulative causation, its performance may decline further. Corporations in countries that become technologically uncompetitive and start to lose market share will see their profits fall, leading to a lower rate of investment, slower technical progress, and hence even greater noncompetitiveness than before. Left to themselves these dynamic market forces can therefore lead to a cumulative decline in a country's share of world markets making it thereby more difficult for its economy to operate at full potential. To counteract such vicious-circle dynamics requires supply-side measures that can improve a country's technological capabilities.

This point can also be looked at from another perspective. What are the implications of technical change *abroad* for a country's standard of living? To analyze this issue, let us suppose that one of the United States' main trading partners (say Japan) has increased its trend rate of technical progress due, for example, to an acceleration in the rate of investment in higher education and science. This has no immediate effect on U.S. productivity growth. Suppose further, however, that this technical change in Japan, although to some extent complementary in the sense of increasing the demand for U.S. goods in Japan, is largely "competitive," that is, Japan starts to export better quality goods in those industries where it competes with the United States. The end result of this process through cumulative causation may be a further decline in U.S. competitiveness and hence productivity growth in the way outlined above, unless the United States is able to adapt to new circumstances either by imitation or by technological development of its own. In the real world of international competition today, such adaptation is crucial to maintain a mix of exports for which the world income elasticity of demand and the potential for productivity growth are sufficient to encourage a virtuous circle of growth, investment, and technical change.

To sum up, contrary to Krugman, there are good analytical and empirical reasons for the view that relative productivity growth does matter profoundly. With so much trade based on nonprice competitiveness, the trade balance can rarely be achieved solely through exchange rate manipulation or only at great cost in terms of employment and real income growth.[7] Moreover, greater productivity growth abroad, as a result of faster technical progress there, is likely to have negative consequences for productivity growth in the home economy unless corrective measures are taken to enhance the country's technological capabilities. Thus even an advanced country cannot afford to ignore its international competitive position if it wishes to improve its standard of living in the long run.[8]

Turning to Krugman's normative objections to the concept of international competitiveness as carrying the implication that trade is a zero-sum game, the foregoing analysis suggests that this objection is also not well founded. To the extent that international competitiveness requires leapfrogging competition between countries in technological developments, this may result in a Pareto improvement in world welfare.

To sum up, once the severe limitations of Krugman's model in its application to the real world are recognized, his analytical and empirical critique of the concept of national competitiveness loses much of its force.

National Competitiveness and U.K. Economic Performance

For much of the post–World War II period the performance of the U.K. economy has been a source of dissatisfaction to economists and policymakers. Although the country recorded faster growth in the post–World War II period than ever before, its relative position declined vis-à-vis other European countries. In the early 1950s, most West European nations had lower productivity levels in manufacturing as well as overall, some very considerably lower than the United Kingdom's. However, by 1987, many of these countries had either caught up with or surpassed the United Kingdom (Maddison 1991, table C11).[9]

There is a large body of literature that suggests that the inadequate performance of the U.K. economy for much of the post–World War II period has been due to the lack of competitiveness of U.K. manufactures in the world economy. There is considerable evidence of such uncompetitiveness. For example, as U.K. manufacturing exports grew at a rate well below that of other advanced industrial countries for much of the postwar period, the U.K. share of world exports declined from over 25 percent in 1950 to just over 9 percent in 1990. Over the same period, Germany's share rose from 7.3 to 20 percent and Japan's from 3.4 to 17 percent. The French share of world exports of manufactured goods remained surprisingly constant—9.9 percent in 1950 and the same in 1990. Like the United Kingdom, the United States also lost share in world markets but at a much slower rate than the United Kingdom. The U.S. share of the world manufacturing exports fell from 27.3 percent in 1950 to 16.2 percent in 1990. Remarkably, this poor performance of U.K. industry in the world economy occurred despite the fact that in the 1960s and 1970s British costs per unit of output expressed in a common currency fell significantly relative to those of other countries.[10]

In analyzing this relative economic decline, following the work of Kaldor, Cambridge economists provided a conceptualization of national competitiveness for the U.K. economy.[11] In a series of papers, Singh (1977, 1979, 1986, 1987, 1989) argued that Britain's poor overall economic performance was due to the failure of its industry in the world economy. He suggested that Britain's industry was "inefficient," and as a consequence of competition from other countries (both advanced and industrializing), it was becoming progressively more so over time. Singh defined an "efficient" manufacturing sector for U.K. economy in the following terms:

> Given the normal levels of the other components of the balance of payments, an efficient manufacturing industry is one which not only meets the needs of the consumers at the lowest cost, but also generates sufficient net exports to pay for the country's required level of imports *at socially desired rates of employment, output growth, and exchange rate, both in the short and long runs.*[12]

In this conceptualization, the qualifications at the end are highly significant. This is because at a low enough level of employment or output, any manufacturing sector would be able to meet this definition of efficiency. However, the question is whether the U.K. manufacturing industry can do so, say at the full employment level of output. Similarly, industry may be able to fulfill these efficiency criteria in the short run as a result of a temporary favorable economic shock (e.g., the discovery of North Sea Oil), but it may be unable to do so in the long run (when the North Sea Oil has run out). The desired level of exchange rate in the above formulation is a surrogate for the socially acceptable rates of inflation and income distribution.[13]

Singh provided evidence to indicate that in these terms the U.K. industry in the 1960s and 1970s was not only inefficient but becoming increasingly so. This was despite the fact that the U.K. costs and prices in common currency were falling rather than increasing relative to those of other countries. As a consequence, the economy, as a whole, was progressively unable to operate at its full potential. Very briefly, the United Kingdom had unfavorable import and export elasticities that indicated over time a growing current account deficit at full employment. In other words, in the international economic regime operating at that time, the country was able to reach a sustainable current account position only at ever-increasing levels of unemployment. Detailed analysis suggested that the main cause of this current account disequilibrium was the poor performance of U.K. industry in the world economy, rather than the competitive failure of other sectors such as agriculture or services.[14]

In the period since 1980 there have been three important changes in the U.K. economy that could in principle alter the above analysis: (1) the discovery of North Sea Oil; (2) the new economic policies of Margaret Thatcher, which differed fundamentally from those adopted by both political parties over the post–World War II period; (3) the new international economic regime of more or less free capital movements, following the abolition of exchange controls in the United Kingdom in 1979.

There is a vigorous debate on the question whether these new elements have made the U.K. industry permanently more "efficient," that is, more competitive. The proponents of the improved efficiency thesis point to the industry's superior productivity growth record in the 1980s and 1990s relative to that of most of the United Kingdom's European neighbors. The opponents point out that the United Kingdom has experienced much steeper deindustrialization than most other industrial countries; that its rate of growth of manufacturing production has also been much slower than that of others; that its comparative investment performance has been very poor; and that there has been very little change in the country's unfavorable propensities to import and export manufactured products.

Whatever one's view about the success (or failure) of the Thatcher measures in rejuvenating U.K. industry, it is clear from the preceding analysis that contrary to Krugman there does exist a meaningful concept of national competitiveness for an advanced country such as the United Kingdom. Such a concept yields important analytical insights. It also has important policy implications that for reasons of space will not be discussed here; interested readers may refer to the large literature on the subject.[15]

Competitiveness and the U.S. Economy

Our analysis, applied to the United Kingdom, illustrated that with an increasingly uncompetitive manufacturing sector in the 1960s and 1970s, and domestic wage-price inflexibility, the United Kingdom became progressively balance-of-payments constrained. The most notable consequence was slow growth and high unemployment.

At first consideration, the U.S. picture appears to be quite different. Despite the fact that between 1978 and 1997 the United States grew at an average annual rate of only 2.3 percent, (roughly comparable to that of Europe), it has experienced much higher rates of employment growth. As a consequence, unemployment rates in the United States have been persistently lower than those in Europe and have recently fallen to levels below 5 percent not seen since the 1960s. Nor have 1960s-style rates of unemployment triggered 1960s-style rates of inflation. More recently, the United States has experienced higher than expected rates of output and productivity growth. The apparent productivity renaissance has triggered a debate about whether the United States is not now finally enjoying the benefits of a "new economy" in which computer technology and increased global competition have relaxed the economic constraints of the 1970s and 1980s (Cairncross 1997a, 1997b; Madrick 1998; Shephard 1997; Tyson 1998).[16]

Does this apparent good performance suggest that the U.S. economy does not suffer from a competitiveness problem? We saw in section 2 that Krugman rejects the idea of a competitiveness problem on the ground that, although the U.S. economy has become much more open than before, the effect of foreign trade on the U.S. standard of living (as mediated through movements in the terms of trade) is too small to count. Moreover, even others who are normally sympathetic to the competitiveness thesis (and whom Krugman criticizes) have been fairly silent in the face of this good economic performance.

There are however important weaknesses in this apparently rosy economic picture, which suggest that attention to competitiveness issues for the U.S. economy may not be entirely out of place. A major blemish in the U.S. economic record is that, although labor markets appear to be tight and there are high levels of employment, real wages have hardly risen at all during the last 25 years. Both productivity growth and per capita income growth, though not as slow as wage growth, have been well below those of the major OECD countries (Howes and Singh 1995). This has meant that the normal expectation of the American people, that each generation's standard of living will be twice that of the previous one, is no longer being realized.

We argue that the main reason for the stagnation of real wages in the United States, like the high rates of unemployment in Europe, is that the economy in the post-1973 period has been expanding at a lower-long term rate than before.[17] Between 1960 and 1973 U.S. GDP grew at an annual rate of 4.0 percent, compared with 2.3 percent since 1973. The higher growth rate of the earlier period not only enabled the country to have a better employment record than it has had subsequently; more significantly, it also made it possible for real wages to increase at a rate of about 2.0 percent per annum in that period.[18]

Thus to meet the historic aspirations of U.S. citizens, it is not enough for the economy to generate high levels of employment, *it must do so with growing real wages.* This can, however, only be accomplished if there is a trend increase in the post-1973 long-term growth rate of the U.S. economy, to the rates that were experienced in the 1950s and 1960s.

An important question is, can the U.S. economy again today expand at the rate of about 4 percent per annum that it achieved in the pre-1973 period?[19] Mainstream economists are fairly united that since about 1973, the maximum possible trend rate of growth of GDP has fallen in the range of 2.0 to 2.5 percent per year. By definition, labor force and productivity growth rates, which are both now averaging about 1.1 percent per year, define the limits to growth. Unlike Europe, where unemployment rates are still multiples above 1960s levels, the fact that unemployment in the United States is now close to levels experienced in the 1960s is taken as a sign that the United States has reached the supply-side limits to its growth rate (Blinder 1997; CEA 1997; Tyson 1998).

This "limits to growth" argument depends crucially on three assumptions. The first is that unemployment rates cannot go much below their present levels (if at all) without setting off inflation—the NAIRU argument. In other words, we are now at full employment, defined as the employment level beyond which the inflation rate will begin to accelerate. The second assumption is that the labor force can grow no faster than its current rate of 1.1 percent per year. The third assumption is that the current 1.1 percent trend rate of productivity growth, which has prevailed for 25 years, is the best that can be expected. Yet all three assumptions are open to question.

Recently, several prominent economists have begun to question the level, mechanism, and consequences of NAIRU. Akerlof, Dickens, and Perry (1996) set NAIRU at 5.0 percent, well below the 6.5 percent rate that dominated the literature until recently. Some (Eisner 1995; Stiglitz 1997) find evidence that the Phillips curve may be concave, implying that even if unemployment is held low for a sustained period, the rate of inflation does not increase at an increasing rate. Although Akerlof et al. find the Phillips curve to be the usual convex shape, Gordon finds the short-run Phillips curve to be resolutely linear. Gordon (1997) estimates that inflation would rise at a steady rate of 0.3 to 0.5 points per year if unemployment were held 1 point below the NAIRU level. Galbraith (1997) argues that if Gordon and Akerlof et al. are right, the inflation costs of holding unemployment 1 point below NAIRU at 4 percent for a decade would be a final inflation rate of 6 percent. But equally significantly, Galbraith has argued historical evidence suggests that the major risks of accelerating inflation have come, not from low unemployment, but rather from external supply-side shocks. Appropriate anti-inflation policy would, under these conditions, be a set of circuit breakers for shock episodes, not slow growth.

But, of course, even if NAIRU is not the binding limit on growth that had previously been supposed, ultimately, without either a faster growing labor force or faster productivity growth, or both, the economy will run up against a supply-side growth constraint. It may be, however, that the potential trend rate of growth of the labor force exceeds the actual or measured trend rate of growth. Recently, Thurow (1996) has argued that approximately one-third of the American work force is looking for more work than it has. In other words, a labor force that is officially 95 percent employed in 1998 is not nearly as employed as a labor force that was 95 percent employed in 1965. And as Bluestone and Harrison (1997) observe, the slightly higher rates of growth in the last few years have revived the upward trend in labor force participation rates, begun to bring back some of the 5 to 6 million young men who had disappeared from labor force statistics (but not from Census statistics) and pushed up the average hours being worked by the typical American worker. So while Blinder (1997) concludes from the falling rate of unemployment that the United States is growing faster than its limit, Bluestone and Harrison take this as evidence, since it has not led to inflation, that "there is a good deal of labor supply in the pipeline when labor demand exists to employ it" (1997, 67). They estimate that an additional 0.3 to 0.4 percent rate of growth can be sustained in the labor force over the next decade.

While there seems to be strong evidence that the labor force is more elastic than has been imagined, in the end, it is significant improvements in productivity growth on which we must rely to raise the trend rate of growth as well as to obtain the required improvement in the growth of real wages. Whether that is possible is a matter of considerable controversy.

Many historians of technology, as well as proponents of the "new economy," regard the cluster of innovations connected with information and communication technology (ICT) to be at par with the two or three most important technical revolutions of the last 200 years, such as steam power, railroads, and electricity (Cairncross 1997a, 1997b; David 1991; Freeman and Soete 1994). However, ICT differs from these in one important respect. In addition to being an input and facilitating production in a wide range of industries, ICT also has direct outputs, in the form of new products, that is, the Internet, CD-ROM, microprocessors and so on (Freeman, Soete, and Efendioglu 1995).

Another important difference between ICT and electricity and steam power is that the pace of technical change in the former has been much faster, with the result that its price has fallen far more quickly relative to the other two. Whereas it took almost 50 years for the price of electricity or of steam power to be halved after the beginning of their commercial use, the price of ICT has already fallen to a fiftieth of what it was 25 years ago. An ordinary PC today costing about $2,000 has as much computing power as the most advanced computer in 1975, costing at the time over $10 million (Woodall 1996).

Yet, despite all this potential, and despite the fact that investment in computers has grown by 30 percent a year on average for the last 20 years, Solow has observed that the computer revolution shows up everywhere but in the productivity data. One explanation for this phenomenon (the so-called Solow paradox) is the view advanced by economic historians that there is always a lag in productivity growth while the new technology is being put in place, while people are learning how to use it (David 1991).

Some have suggested that benefits of the new technological revolution are already occurring but are simply being mismeasured. The Boskin report, for example, found that the current method of measuring the CPI overstated the rate of inflation by approximately 1 percentage point. This finding was used to argue—and this was perhaps one of the motivations for attacking the CPI in the first place (Baker 1998)—that the real rate of growth of output and productivity was being understated.[20] Mismeasurement of the CPI, however, offers little explanation for the productivity slowdown (Solow 1998). As even members of the Commission have acknowledged, if the CPI is overstated, it was overstated both before and after 1973.

More recently Solow (1998) and Madrick (1998), citing the work of Sichel (1997), reject the view that there is any productivity resurgence, either apparent or hidden in inadequately measured data. Using growth accounting methods to measure the contribution of computer investment to output and productivity growth, Sichel finds that because computers still represent a very small share of total capital stock, their contribution to output growth could not have been large. Only if computer investment earned a rate of return far in excess of the normal rate of return—in which case firms would be irrationally underinvesting—could the contribution of computers be significantly larger.[21]

What is missing from this whole discussion is the effect that slow growth has had on the realization of the potential of the new technologies. The growth accounting framework generally abstracts from the role of demand. However, if the rate of growth of *real* aggregate demand were higher, industries would have made faster progress putting into place all the pieces necessary—software, hardware, and skills—to achieve the full potential.[22] ICT would have more widespread use in most branches of industry and services, reducing their prices and, as in the case of previous technical revolutions, leading to a virtuous circle of increased demand, increased output, and increased growth of productivity.

In short, to realize the supply-side potential of the ICT revolution for the economy as a whole, that is, to raise the trend rate of growth of productivity, a faster rate of growth of real demand is needed. This in turn would lead to fast growth of output as well as of the labor force. Were the trend rates of growth of the labor force and productivity to rise, whatever real inflation constraint exists would be relaxed.[23]

However, even if the inflation constraint could be overcome, it is arguable that the United States is already currently growing close to its maximum sustainable rate, given its uncompetitive manufacturing sector. The U.S. current account deficit would, other things being equal, greatly increase with the expansion of real aggregate demand. The deficit would most likely become unsustainable.

Analysis of trends in the current account reveals that, since 1978, there has been deterioration in the current account for a given rate of growth. The deterioration is due primarily to unfavorable trends in the income elasticities of demand for the country's imports and exports. The fact that the rate of growth of some of the United States' primary trading partners has slowed more than that of the United States has also contributed to the trade imbalance. And the failure of the U.S. currency to continue to depreciate as it did in the 1960s and 1970s has meant that the relative price position has worsened as well.

From about 1960 to 1973, the United States was able to sustain average annual rates of growth of 4 percent and maintain a small positive balance on the current account, albeit with a steadily depreciating real exchange rate and relative unit labor costs. During this period, it should be noted that world GDP was growing about 5 percent a year, 1 percentage point faster than the United States. Between 1973 and 1979, United States growth rates slowed to about a 2.8 percent average, while the rest of the world continued to grow at close to 5 percent a year. The U.S. was still able to sustain a positive current account balance but only with the real exchange rate depreciating at about 3 percent annually, similar to the average rate of depreciation between 1960 and 1973.

Since 1980, the United States has run a deficit on the current account that has averaged about 1.5 percent of GDP. Two changes have affected the ability of the United States to maintain balance in the current account. First, while over the entire period, the rest of the world has continued to grow faster than the United States the growth gap has narrowed substantially. There have been a couple of periods when the rate of growth of demand in the United States has actually outstripped that of the rest of the world. Second, the real value of the dollar reversed its long-term decline, starting in 1980. In real terms, the value of the dollar did not fall, on average, from 1980 to 1995, though, as is well known, there was a stretch between 1979 and 1985 when the dollar appreciated substantially, falling back to its 1980 value by 1987. Since 1995 the dollar has once again begun to appreciate (Klitgaard and Orr 1998).

It is certainly true that the largest current account deficits have been experienced during periods of exceptionally slow relative growth for the rest of the world, or during periods of substantial appreciation of the dollar. However, it is not the case that these exceptional circumstances fully explain the persistent current account deficit, for even in periods of relative rapid growth for the rest of the world, or when the U.S. dollar has not been appreciating, the deficit has persisted. Blecker (chap. 2, this vol.) points out that, if instead of not depreciating at all between 1980 and 1993, the dollar had continued to depreciate at the trend rate of about 3 percent a year that had prevailed in the 1970s and which was then sufficient to balance the current account, by 1993, the dollar's value would have been 40 percent below what it actually was.

A simple simulation estimating the impact of relative rates of growth and currency depreciation on the U.S. current account, given current income and price elasticities for imports and exports, renders the following scenarios.[24]

Suppose that we accept the current 3 percent rate of growth of the rest of the world as given. Then if the United States wants to see its current account deficit grow no faster than the rate of growth of GDP, it has three options: (1) grow at its current long-term rate of 2.3 percent and allow its currency to depreciate at an annual rate of a little over 1 percent a year; (2) grow at a slower rate of 1.3 percent without any currency depreciation; or (3) grow at the faster rate of the 1960s—4 percent—and allow its currency to depreciate each year by 3 percent in real terms.

If the United States continues to grow at its current 20-year average of about 2.3 percent, while the rest of the world continues to grow at its 20-year average of 3 percent, unless the U.S. currency depreciates, the current account deficit as a percent of GDP will continue to grow. Within 10 years it will be about 5 percent of GDP. If the United States tries to grow faster, say at the 4 percent rate that prevailed in the 1960s, given the slow rate of growth of the rest of the world, the current account deficit would reach about 10 percent within 10 years. Of course, if the U.S. currency continues to appreciate, as it has for the last 2 years, the deficits will be larger.[25]

A current account deficit of this magnitude would certainly be difficult to finance, even for the U.S. economy, and even under currently prevailing conditions of free international capital flows. To attract this volume of foreign debt would require high interest rates, which would slow the rate of economic growth. Further, Godley and Milberg (1994) and Howes (this vol.) have shown that, even if the United States could continue to finance the debt, the total external debt in ten years at projected deficit levels would be unsustainable.[26]

The foregoing analysis suggests that the United States faces an exceedingly unattractive menu of policy options—some combination of currency depreciation, high interest rates, and slow growth—in order to meet the balance of payments constraint on faster growth required to enable the realization of full employment with growing real wages. Of course, the United States could also achieve a better current account position without resorting to any of these measures if other countries simply grew faster than their current rates. Such a scenario seems highly unlikely, however, in the current circumstances of the global economy. For the global economy to grow at a substantially faster rate would require a high degree of cooperation among the industrial countries. In particular it would require that both Germany and Japan sustain sufficiently high rates of growth to balance their external accounts, a path neither seems able or inclined to follow.

Another option would be for the United States to improve the competitiveness of its export sector, especially that of manufactured goods.[27] People take different views on how greater competitiveness might be achieved. What little favor was held by industrial policy, including both protection measures and subsidies to strategic industries, has generally been supplanted by competition policy due both to ideological reasons and to the restrictions imposed by membership in the World Trade Organization. Also in current favor are proposals to increase the savings and investment rate at current rates of growth.

However, given the present state of the world economy, were the United States to increase the competitiveness of its manufacturing sector without increasing its overall rate of economic growth, other things being equal, this would have the equivalent impact on world growth of imposing import restrictions. A more competitive U.S. manufacturing industry will reduce the trading partners' exports and increase their imports. The rate of growth of our major trading partners, including Japan and other Asian countries, would be slowed as a consequence. Therefore, in the interests of the long-run vitality, of both the U.S. and the world economy, it is necessary for the United States to increase both the competitiveness of its exports and its rate of growth simultaneously. In other words, the United States must increase the propensity to export and reduce the propensity to import at a given rate of growth while simultaneously increasing the rate of growth. In such a scenario, U.S. action would stimulate the rate of growth of the rest of the world because the volume of its imports would actually rise, even while its propensity to import fell.

In light of the foregoing analysis, the Asian financial crisis raises the following questions with respect to the long-term prospects for sustainable rates of growth in the United States. The huge depreciation of Asian currencies since spring of 1997 has put upward pressure on the already appreciating United States dollar. Since late 1994, the real effective exchange rate for the U.S. has, by some measures, appreciated by over 20 percent (Klitgaard and Orr 1998). The currency realignments combined with declining growth rates of the Asian countries have led the OECD to predict a cumulative positive current account adjustment for Japan and emerging Asia of $113 billion dollars in 1998 and 1999. The U.S. current account is projected to experience an $80 billion adjustment, $50 billion from emerging Asia alone, not including Japan (OECD 1998). In the current context of unstable financial markets, such an increase in the deficit may not be sustainable. This would lead to either a depreciation of the U.S. currency or slower growth. A better option would be for the United States to improve the competitiveness of its manufacturing industry, while sustaining high rates of overall economic growth. With the approach of a presidential election year, there is bound to be pressure for protection in Congress, most likely in the form of trade restrictions on countries found to be guilty of human rights violations.

At the same time, the Asian financial crisis does reduce the threat of inflation transmitted through oil and commodity prices as well as imports of manufactured goods from Asia. Therefore, the above analysis suggests that the best approach for the United States, and for the global economy in general, would be to absorb the imports from the emerging Asian economies as well as support massive loans from the IMF and other international agencies, so that Asian economies are able to finance their recovery. The financial markets are more likely to accept this greater U.S. deficit if it does not rise as a proportion of GDP, and if there is visible improvement in the competitiveness of the U.S. tradable goods sector. At the same time, given the reduced threat of inflation (whether real or political), now is a good time for the United States to maintain the high rates of growth necessary to encourage investment in domestic industries.

In this context, the U.S. Federal Reserve's recent reassessment of the balance of risk between deflation and inflation leading to expansionary policies is to be welcomed. However, such policies need to be intensified. But to do so, they must be accompanied by measures that will either reduce U.S. consumption and/or increase net exports for a given growth rate along the lines suggested above.

The main thesis of this introduction is that both from the short-term and long-term perspectives of the United States as well as the world economy, the United States needs to improve its national competitiveness in order to achieve sustained faster growth of output and productivity. Opinions will differ, as indeed they do in this book, on how national competitiveness may best be enhanced—whether through improvements in export and import propensities, and/or through an increase in the domestic savings rate. The amelioration of the competitiveness constraint and faster economic growth in the United States will not only help raise employment and real wages in the U.S. economy but will also aid world economic recovery in the short term. This analytical and policy perspective is also in the long-term interest of both the U.S. and the world economies.

The remaining sections of this book focus on analytical, empirical, and policy issues related to competitiveness of the U.S. economy. Part 1 explores the reasons for the deterioration in the trade deficit in the 1980s. Part 2 is concerned with the effect of the institutional arrangements in financial markets on both price and especially nonprice competitiveness. Part 3 examines past and present technology policies, and part 4 is concerned with issues of industrial policy.

In part 1 of the book, Catherine Mann and Robert Blecker each analyze the trend in the U.S. trade deficit over the 1980s, including the extent to which it may be related to deterioration in competitiveness, narrowly defined as price competitiveness. Mann offers a simple analytical model of firm behavior to illustrate how macroeconomic and microeconomic policies can affect international price competitiveness. Using this framework, she analyzes the effect of trends in relative prices, relative income growth, and exchange rates on the trade deficit and draws policy implications for improving international competitiveness.

In addition to his analysis of lessons to be drawn from the trade deficit of the 1980s, Blecker also provides an important analytical argument for why structural and macroeconomic explanations of the trade deficit are not mutually exclusive. He shows analytically that once the assumption of full employment in the general equilibrium model is dropped, it is possible for changes in the relative competitiveness of a nation's tradable goods simultaneously to affect both savings and investment, on the one hand, and the trade balance, on the other. He also shows that improvements in competitiveness will not necessarily be offset by an appreciation in the exchange rate.

Both Mann and Blecker find that while controlling for trends in relative growth rates and prices, the trend in the trade deficit between 1980 shows a deterioration, and to that extent, the deficit is structural. Mann suggests that unfavorable differential rates of productivity growth between countries may explain the trend. Blecker attributes the U.S. trade deficit to structural problems primarily with Japan and the newly industrializing Asian countries.

While Blecker and Mann reach similar conclusions about the long-run trend in competitiveness for U.S. manufacturers, they recommend very different remedies. From her model, Mann argues that any improvements in price competitiveness must be effected through growth differentials, relative prices, or the exchange rate. During the period under consideration, exchange rate movements have outweighed movements in any other determinant of price competitiveness. She regards macroeconomic policy management to be the key variable. Within her framework of analysis, she finds no important mechanism by which managed trade could affect any of the variables relevant to price competitiveness. Productivity growth, she argues, is also key if the United States is to avoid having to close the external gap through a policy of continual currency depreciation; faster productivity growth is generated by "a stable macroeconomic environment complemented by flexible and competitive markets."

Blecker, on the other hand, suggests the U.S. trade deficit primarily with Japan and the newly industrializing Asian economies cannot be corrected by normal market processes. These imbalances can be remedied by a combination of both macroeconomic coordination and managed trade.

The chapters in part 2 on finance and international competitiveness complement each other. Lazonick and O'Sullivan's chapter traces the historical evolution of the U.S. system of corporate governance from the inception of a market for industrial securities in the 1890s to the present state of development of the market for corporate control. They make the important point that it was not the development of the U.S. securities market that led to the expansion of large U.S. corporations but the other way around: the expansion of large corporations led to the development of the U.S. securities market. The U.S. securities market development was a by-product of the large U.S. merger and trust movement of the turn of the century. They ascribe the American industrial ascendancy during the first half of the century to the separation of ownership and control that allowed professional managers to run these corporations as organizational entities that facilitated innovation and productivity growth. The development of the market for corporate control led, in the second half of the century, to operational specialization and the dominance of finance over production activities within the organization. Such a system may be useful for creating shareholder value, but they argue that it is not conducive to advancing technical progress and international competitiveness.

Singh's paper explores the links between the market for corporate control and international competitiveness. Central to this analysis is the overall rate of investment and its time horizon. The higher the overall rate of investment, the greater the turnover of machines, the faster the technical progress and the quicker the development of new products, over time, the greater will be international competitiveness. He particularly emphasizes the need for incentives for both firms and their employees to acquire firm-specific human and organizational capital. For various reasons, the market for corporate control leads to short-termism and blunts the incentive for acquisition of firm-specific human capital. At the aggregate level, short-termism may also contribute to raising the cost of capital. The paper puts forward proposals for a system of corporate governance that, by throwing sand in the gears of the market for corporate control, restrains short-termism.

The two chapters by Markusen and Steinmueller in part 3 are concerned with technology policy. Both authors view technology policy as a component and/or complement to industrial policy. Markusen advocates the use of technology policy to further the long-term goal of high-wage job creation. She takes particular exception with "competitiveness" as a driving force for technology policy, arguing that winner-take-all competitive contests among advanced industrial nations could generate a simultaneous decline in general levels of income. Steinmueller argues that technology policy should be aimed at technologically progressive industries, those that either produce or bring into use new technologies that enhance productivity or stimulate economic growth.

Markusen's chapter provides an overview of the strengths and weaknesses of technology policy as currently proposed for the United States. In so doing, she assesses past efforts to evaluate the successes and failures of technology policy and concludes that there is much still to be understood before we base current technology policy on past practice. That said, she draws lessons from the past as the basis for three speculative arguments. First, without complementary policies aimed at market creation and favorable finance and trade environments, the supply-side focus on the implementation of technology policies is unlikely to succeed. Second, while the subsidization of new technologies is defended as a means of long-term creation of good jobs in the American economy, the institutional arrangements under which subsidization is being pursued do not ensure that gains will translate into domestic job generation. In fact, she argues, it is possible that subsidization of new technologies may result in disruptive and unbalanced growth dynamics, accelerating labor-displacing innovations without creating absorptive capacity elsewhere in the economy. She suggests that the focus on high technology policies may lead to a worsening of income distribution. A way out of the dilemma, she argues, is to harness technology policy to new missions in the economy that are addressed to pressing national needs on, for example, the environmental, energy, and transportation fronts. She concludes that no irreversible decisions regarding the size and institutional form of technology policy should be undertaken at this point without more public debate about the options and better evaluation of the past record.

Steinmueller turns also to past experience with technology policy to situate his argument. He argues, following Abramowitz, that while U.S. relative industrial decline is due to expenditures on Cold War and convergence, we must still account for the 25 years that the United States was able to retain technological leadership. He concludes, following Nelson, that leadership resulted from a set of policy decisions that enhanced U.S. ability beyond that of other industrial nations, including policies to raise the share of college-educated in the work force, funding for university research, and applied research efforts of the DOD. Lessons from the past suggest that technological and industrial leadership could be recovered by investment in human capital formation, national infrastructure, development of technologically intensive national missions, and government policies designed to create new types of demand.

Part 4 contains two chapters that are concerned with the need for industrial policy in the United States. Howes provides the full empirical argument, summarized earlier in this introduction, that U.S. growth is balance of payments constrained. She concludes that U.S. competitiveness problems require remedial policies for industries strategic to achieving balance on the external account. Luria's chapter provides a detailed argument for why the United States needs an industrial policy to encourage small manufacturing firms to upgrade their production technologies and the skills of their workers. While Luria's chapter stresses the role played by small businesses in the productivity growth slowdown, and while his recommendations appear to be sectorally neutral, he shows that, in fact, the firms most likely to be affected are in industries and regions that have historically accounted for a large share of net exports. Thus, his policy recommendations may provide a practical solution to the competitiveness problems that confound U.S. efforts to grow faster.

Luria's paper considers first the contribution made to the U.S. productivity slowdown by the redistribution of manufacturing production over the last 25 years from large to small plants. Not only do small plants have lower productivity (and wages) than large plants, but the gap has been increasing as the distribution has shifted toward small plant production. A similar shift to small plant production in Europe and Japan, Luria argues, has not led to a widening in the productivity gap between small and large plants. Luria attributes the relatively better performance of small European and Japanese plants to a regulatory environment, including policies to promote wage equalization between small and large plants, that biases small-firm production strategies toward a high-wage, high-productivity option. Luria suggests a set of policies, first and foremost including robust economic growth, that would change the incentives to firms facing a choice between low and high technology production strategies.

Interest in competitiveness issues waxes and wanes with the U.S. business cycle and its position relative to other countries. After the cyclical downturn of the early 1990s, the rest of the decade has seen an unprecedented prosperity with the longest postwar boom. Europe and Japan, on the other hand, have suffered stagnation relative to past experience. However, as Europe and Japan overcome their current difficulties and recover their previous trend rates of growth, the competitiveness issues of the U.S. economy will again loom large in the public eye.

NOTES

1. Krugman suggests that such economists know that national competitiveness is a meaningless term, but may believe such a rhetoric of competitiveness to nevertheless be useful to the extent that it leads to greater expenditure on good things like education.

2. As noted above, according to Krugman, terms-of-trade adjustments have, in reality, little impact on real incomes.

3. See further Singh 1987.

4. See Fagerberg 1996; see also Fagerberg 1988, for the original extension of Kaldor's work. McCombie and Thirlwall (1994) survey a number of empirical studies, including Fagerberg 1988, all of which confirm that the growth of import and export market shares is not significantly affected by changes in relative prices. It should be noted that it is customary to examine the relationship between export shares and relative price competitiveness using RULC rather than relative export prices. This is because relative export prices are measured by dividing the total value of exports by units, and so rising relative export prices may simply be capturing the effect of improvements in quality and other nonprice attributes (Kaldor 1978).

5. While a conventional view would argue that the causal link between productivity growth and market share was through the effect of productivity growth on price competitiveness, the fact that relative unit labor costs are rising with productivity and market share, rather than falling, would seem to contradict that causality, at least for countries that export advanced industrial products. Kaldor explains that both relative unit labor costs and productivity are rising because the more competitive a nation's exports become, based on nonprice factors, the faster will rise that nation's world export market share, simultaneously pulling up its exchange rate and relative export prices. Hence, Kaldor concluded that causality ran, not from relative price to market share, but from market share to price. The change in "competitiveness," as conventionally measured by relative prices and unit labor costs, was not the cause but the consequence of the change in market shares. The underlying trends in market share, as Kaldor argued, must be due to nonprice factors not susceptible to measurement.

6. For a fuller discussion of the relationship between technology and exports, see reviews by Dosi, et al. (1990), Fagerberg (1996), and McCombie and Thirlwall (1994).

7. Dornbusch (1996) shows that real exchange rate adjustments are still possible but only when there are high rates of unemployment.

8. See Dowrick 1997 for a recent review of the voluminous literature on the two-way causal relationship between foreign trade and economic growth.

9. Krugman's suggestion that relative productivity does not matter and that only absolute productivity is important is simply not a valid statement for the U.K. economy for yet another significant reason, in addition to those outlined in the previous section. The leadership of the leading political parties as well as the British public were very conscious of the fact that although they were better off than before, they were falling behind their European neighbors. There was general agreement that this relative decline should be arrested.

10. Sources: National Institute Economic Review, various issues; Brown and Sheriff 1979, Singh 1977. The "world" is defined as the following countries: the United Kingdom, France, Germany, Italy, Japan, the United States, Canada, Sweden, Switzerland, and the Benelux countries.

11. These issues were discussed in the context of the debate on the U.K. deindustrialization.

12. The reader will note the broad similarity between this definition of national competitiveness and that provided by Tyson (1992) for the U.S. economy.

13. For a detailed commentary on this definition of an efficient manufacturing sector for the U.K. economy, see Cairncross 1979. Cairncross christened this the Cambridge view of U.K. deindustrialization.

14. It is important to note, in light of Ball 1990, Ball and Robertson 1993, and Krugman 1989, that we are not here arguing that income elasticities of demand for imports and exports are constant. Ball (1990) first argued, in a discussion of Singh 1977, that the balance-of-payments constraint theory depended on the assumption that there was a single full employment level of imports, such that both internal and external balance could not simultaneously be achieved. Ball and Robertson (1993) revisit the issue in a lengthy critique of Thirlwall 1992, in which they invoke Krugman's (1989) claim that over the long term income elasticities cannot account for differences in rates of growth. Again they argue that the balance-of-payments constraint theory requires the assumption that income elasticities of demand for imports and exports are constant.

What we do argue, as did Singh (1977, 1989), and Kaldor before him, is that whether income elasticities are large or small is a matter of the innovative and adaptive capacity of the nation's entrepreneurs. In other words, unfavorable income elasticities are indeed a supply-side problem. However, once the poor performance of the manufacturing sector began to manifest itself as a balance of payments problem, the low demand for British exports further depressed investment and innovation so that through cumulative causation the competitiveness of U.K. industry continued to deteriorate.

15. For a recent comprehensive discussion of the competitiveness of the U.K. industry and the policy implications that follow from it, see the symposium in *The Economic Journal,* including Crafts 1996 and Kitson and Michie 1996.

16. Madrick (1998) challenges the idea that higher rates of productivity and output growth for 1996 and 1997 reflect any sort of economic renaissance. As he shows, despite even the rapid growth since 1996, the expansion of the economy that began in 1990 is the slowest expansion of the entire post–World War II period. Single spurts of productivity growth, such as have occurred in 1996 and 1997, are hardly unprecedented, nor are they particularly impressive, compared to, say, productivity growth in 1986 or in the late 1970s.

17. Due to the fact that productivity growth is so much higher in Europe than in the United States, the consequence of equally slow rates of growth in the two regions has been high unemployment for Europe and full employment with stagnant real wages for the United States. Given the differences in labor market institutions, to achieve full employment in Europe would require rates of growth comparable to those of the Golden Age in the range of 4 to 5 percent annually. The United States can achieve full employment at lower rates of growth because productivity growth is so low. However, raising standards of living will require higher rates of productivity growth.

18. The average real earnings for nonsupervisory workers in the United States declined at a rate of 0.6 percent per annum in the 1970s, 1 percent per annum in the 1980s, and 0.6 percent per annum between 1990 and 1995. In contrast, in the 1950s and 1960s the corresponding average real earnings rose at a rate of 2.0 percent per annum (Mishel, Bernstein, and Schmitt 1997).

19. As will be argued below, it is important to note here that if this faster growth were possible this would normally not only benefit the U.S. economy, but also the rest of the world.

20. Baker (1998) points out that more recently, proponents of the "new economy" have taken heart in another apparent data problem, the statistical discrepancy between income-based and output-based measures of GDP. They have argued that the income-based measure, which has been growing more rapidly than the output-based measure, is the more accurate. This would imply that output and productivity growth for 1996–97 were considerably higher by that measure. As this controversy is only beginning, we can only report that Baker offers a compelling counterargument, which will have to be refuted.

21. Brynjolfsson and Hitt (1996) do find that the rate of return to computer investment is above the normal rate of return. Using their estimated rates of return, Sichel recalculates the contribution of computers to output and productivity growth and finds it still to be extremely low. But in the end he rejects the idea that rates of return to computer investment could be terribly high because that implies irrational behavior by firms.

22. See Bluestone and Harrison 1997, and Cairncross 1997a,b for the argument

that the full potential of a technology revolution is not realized until all the components are in place.

23. That is not to imply that there will not be a political inflation constraint as long as the Federal Reserve is charged with conducting anti-inflation policy.

24. See Howes in this volume for more detail. These estimates use the most recent calculations of income and price elasticities from Hooper, Johnson, and Marquez 1998. However, it should be noted that the authors also find strong evidence that the income elasticity of demand for U.S. exports experienced a steady decline over the period 1985 through 1995, while the income elasticity of demand for imports showed a slight rise. Clarida and Hickok (1993) offer good evidence for why, due to the declining competitiveness of the U.S. capital goods sector relative to other countries, we should expect to see the income elasticity of demand for U.S. exports fall further while that of imports rises. The capital goods sector is generally considered to have higher income elasticity than other manufactured goods have. Since 1978, the share of this sector in U.S. goods exports has risen 9 points, while its share in imports has risen 18 points. These trends are not driven by differences in relative rates of growth of demand. Because the weight of this sector in imports is rising so much more rapidly than in exports, one should expect to see the income elasticity of imports rising faster than that of exports.

25. See further Howes in this volume. Clearly, these calculations make numerous assumptions. One major abstraction from reality involves the treatment of investment income flows in the current account. Since the income elasticities measure only the effect of income changes on the flow of goods and services, we are here assuming that factor income flows will not offset the trends in goods and services. As the next note suggests, net income flows are likely to worsen the current account deficit.

26. By the end of 1997, net foreign debt for the United States exceeded $1 trillion, or approximately 12 percent of GDP. For the first time ever, net income flows on net foreign assets were definitively negative in 1997; the net outflow was $5 billion dollars. What was surprising was that net income flows had continued to be positive for 10 years despite the fact that the United States had been a net debtor since 1988. As Godley and Milberg (1994) showed, however, this paradox could be explained by the fact that the rate of return on U.S. assets abroad exceeded the rate of return on foreign holdings in the United States, a situation which was unlikely to persist. In fact, rates of return have been converging in the last few years, and it seems possible that before too long the return on foreign assets in the United States will exceed return of U.S. assets abroad. This is due, in large part, to the fact that the mix of assets and liabilities is becoming increasingly weighted to portfolio investment, on which the rates of return for foreign assets in the United States exceed those for U.S. assets abroad. In the meantime, the weight of direct investment, where large differentials in rates of return had existed, has fallen, while the rates of return have simultaneously converged. The final factor that supports the conclusion that net income flows will become increasingly negative is the fact that inward investment has been growing at a rate 2 to 3 percentage points higher than outward investment since at least 1990, and 6 percentage points higher in the last three years. While any significant realignment of interest rates would undoubtedly alter this pattern, the fact that the Federal Reserve seems to believe that the U.S. economy is now growing at its maximum possible rate, while both Europe and Japan are considered to be growing at below their potential rates, does not suggest that U.S. interest rates are likely to fall relative to these regions anytime soon. (See Howes 1999 in this volume for more detail, especially table 9.3.)

27. See Howes (1999) in this volume for a discussion of why services exports cannot compensate for an uncompetitive manufacturing sector.

REFERENCES

Akerlof, George A., William T. Dickens, and George L. Perry. 1996. The macro-economics of low inflation. *Brookings Papers on Economic Activity* 27, no. 1:1–26.

Bach, Christopher L. 1998. U.S. International Transactions, Fourth Quarter and Year 1997. U.S Department of Commerce, *Survey of Current Business* (April): 51–97.

Baily, Martin, and Robert Gordon. 1988. Measurement Issues, the Productivity Slow-down and the Explosion of Computer Power. *Brookings Papers on Economic Activity* 2:347–431.

Baker, Dean. 1998. The New Economy. *Challenge* 41, no. 6: 5–10.

Ball, James. 1990. Manufacturing Industry, Economic Growth and the Balance of Payments. In *Money, Trade and Payments,* ed. D. Cobham, R. Harrington, and G. Zis. Manchester: Manchester University Press.

Ball, James, and Donald Robertson. 1993. Briefing Paper: Manufacturing Industry and the Balance of Payment Adjustment: The United Kingdom Case. *OECD Economic Outlook* 53 (June): 39–49.

Blecker, Robert A. 1992. *Beyond the Twin Deficits: A Trade Strategy for the 1990s.* Armonk, N.Y.: M. E. Sharpe, Inc., Economic Policy Institute Series.

———. 2000. The Trade Deficit and U.S. Competitiveness. In this volume.

Blinder, Alan S. 1997. The Speed Limit: Fact or Fancy in the Growth Debate. *The American Prospect* no. 34 (September/October): 57–62.

Bluestone, Barry, and Bennett Harrison. 1997. Why We Can Grow Faster. *The American Prospect* no. 34 (September/October): 63–74.

Brown, C. J. F. and T. D. Sheriff. 1979. Deindustrialisation: A Background Paper. In *Deindustrialisation,* ed. Frank Blackaby. London: Heinemann Educational Books.

Brynjolfsson, Eric, and Lorin Hitt. 1996. Paradox Lost? Firm-Level Evidence on the Returns to Information Systems Spending. *Management Science* 42 no. 4 (April).

Cairncross, Alec. 1979. What Is Deindustrialisation? In *Deindustrialisation,* ed. Frank Blackaby. London: Heinemann Educational Books.

Cairncross, Frances. 1997a. A Connected World. *The Economist* (September 13, 1997).

———. 1997b. *The Death of Distance: How the Communications Revolution Will Change Our Lives.* Boston: Harvard Business School Press.

Carnevale, Anthony, and Stephen Rose. 1998. *Education for What? The New Office Economy.* Princeton, N.J.: Educational Testing Service.

Clarida, Richard H., and Susan Hickok. 1993. U.S. Manufacturing and the Deindustrialisation Debate. *World Economy* 16 no. 2 (March):173–92.

Cline, William R. 1989. *United States External Adjustment and the World Economy.* Washington, D.C.: Institute for International Economics.

Council of Economic Advisors. 1997. *Economic Report of the President.* Washington, D.C.: U.S. Government Printing Office.

Crafts, Nicholas.1996. Deindustrialisation and Economic Growth. *The Economic Journal* 106 (January): 172–83.

David, Paul. 1991. Computer and Dynamo: The Modern Productivity Paradox in a Not-Too-Distant Mirror. In *OECD, Technology and Productivity: The Challenge for Economic Policy.* Paris: OECD.

Denison, E. 1989. *Estimates of Productivity Change by Industry: An Evaluation and an Alternative.* Washington: The Brookings Institution.

Dornbusch, Rudi. 1996. The Effectiveness of Exchange-Rate Changes. *Oxford Review of Economic Policy* 12, no. 3 (autumn): 26–38.

Dosi, Giovanni, Keith Pavitt, and Luc Soete. 1990. *The Economics of Technical Change and International Trade.* New York: New York University Press.

Dowrick, S. 1997. Trade and Growth: A Survey. In J. Fagerberg et al., *Technology and International Trade.* Aldershot, U.K.: Edward Elgar.

Eisner, Robert. 1995. Our NAIRU Limit. *The American Prospect* 21 (spring): 58–63.

Fagerberg, Jan. 1988. International Competitiveness. *The Economic Journal* 98, no. 2, (June): 355–74.

———. 1996. Technology and Competitiveness. *Oxford Review of Economic Policy* 12, no. 3 (autumn): 39–51.

Freeman, Chris, and Luc Soete. 1994. *Work for All or Mass Unemployment: Computerised Technical Change into the Twenty-First Century.* London: Pinter.

Freeman, Chris, Luc Soete, and Umit Efendioglu. 1995. Diffusion and the Employment Effects of Information and Communication Technology. *International Labour Review* 134, nos. 4–5: 587–603.

Galbraith, James K. 1997. Time to Ditch the NAIRU. *Journal of Economic Perspectives* 11, no. 1: 93–108.

Godley, Wynne, and William Milberg. 1994 U.S. Trade Deficits: The Recovery's Dark Side? *Challenge* 37, no. 6 (November/December): 40–47.

Gordon, Robert J. 1997. The Time-Varying NAIRU and Its Implications for Economic Policy. *Journal of Economic Perspectives* 11, no. 1, (winter): 11–32.

Helkie, William L., and Peter Hooper. 1988. An Empirical Analysis of the External Deficit. In Ralph C. Bryant et al., eds., *External Deficits and the Dollar: The Pit and the Pendulum.* Washington, D.C.: The Brookings Institution.

Hooper, P., and J. Marquez. 1995. Exchange Rates, Prices and External Adjustment in the United States and Japan. In *Understanding Independence,* ed. P. Kenen 107–68, Princeton, N.J.: Princeton University Press.

Hooper, Peter, Karen Johnson, and Jaime Marquez. 1998. Trade Elasticities for G-7 Countries. Board of Governors of the Federal Reserve System, International Finance Discussion Papers, no. 609 (April).

Howes, Candace. 1995. Long Term Economic Strategy and Employment Growth in the U.S.: An Analysis of Clinton's Economic Policies. *Contributions to Political Economy* 14:1–31.

———. 2000. Competitiveness and U.S. Economic Growth. In this volume.

Howes, Candace, and Ajit Singh. 1995. Long-Term Trends in the World Economy: The Gender Dimension. *World Development* 23, no. 11: 1895–1911.

Kaldor, Nicholas. 1978. The Effects of Devaluation on Trade in Manufactures. In N. Kaldor, *Further Essays on Applied Economics.* London: Duckworth.

———. 1981. The Role of Increasing Returns, Technical Progress and Cumulative Causation in the Theory of International Trade and Economic Growth. *Economie appliquée* 34, no. 4: 593–617.

Kitson, Michael, and Jonathan Michie. 1996. Britain's Industrial Performance Since 1960: Underinvestment and Relative Decline. *Economic Journal* 106 (January): 196–212.

Klitgaard, Thomas, and James Orr. 1998. Evaluating the Price Competitiveness of U.S. Exports. Federal Reserve Bank of New York, *Current Issues in Economics and Finance* 4, no. 2 (February): 1–6.

Krugman, Paul. 1989. Differences in Income Elasticities and Trends in Real Exchange Rates. *European Economic Review* 33: 1031–54.

———. 1994. Competitiveness: A Dangerous Obsession. *Foreign Affairs* 73 (March/April): 28–44.

———. 1996. *Pop Internationalism.* Cambridge, Mass.: MIT Press.

Lawrence, Robert Z. 1990. U.S. Current Account Adjustment: An Appraisal. *Brookings Papers on Economic Activity,* no. 2: 343–89.

Maddison, Angus. 1991. *Dynamic Forces in Capitalist Development.* Oxford: Oxford University Press.

Madrick, Jeffrey. 1997. *The End of Affluence.* New York: Random House.

———. 1998. Computers: Waiting for the Revolution. *Challenge* 41, no. 4 (July–August): 42–65.

McCombie, J. S. L., and A. P. Thirlwall. 1994. *Economic Growth and the Balance of Payments Constraint.* London: Macmillan.

Mishel, L. 1988. *Manufacturing Numbers: How Inaccurate Statistics Conceal U.S. Industrial Decline.* Washington: Economic Policy Institute.

———. 1989. The Late Great Debate on Deindustrialization. *Challenge* 32, no. 1 (January): 35–43.

Mishel, Lawrence, Jared Bernstein, and John Schmitt. 1997. *The State of Working America, 1996–97.* Washington, D.C.: The Economic Policy Institute and ME Sharpe.

Nelson, Richard R., and Gavin Wright. 1992. The Rise and Fall of American Technological Leadership: The Postwar Era in Historical Perspective. *Journal of Economic Literature* 30, no. 4 (December): 1931–64.

OECD. 1998. *OECD Economic Outlook,* no. 63, June.

Shephard, Stephen B. 1997. The New Economy: What It Really Means. *Business Week* (November 17).

Sichel, Daniel. 1997. *The Computer Revolution: An Economic Perspective.* Washington, D.C.: The Brookings Institution.

Singh, Ajit. 1977. "UK Industry and the World Economy: A Case of De-industrialization? *Cambridge Journal of Economics* 1, no. 2 (June): 113–36.

———. 1979. North Sea Oil and the Reconstruction of the UK Industry. In *Deindustrialisation,* ed. F. Blackaby. London: Heinemann Educational Books.

———. 1986. Long-term Structural Disequilibrium of the UK Economy: Employment, Trade and Import Controls. In *Free Trade—Managed Trade? Perspectives on a Realistic International Trade Order,* ed. G. Sjostedt and B. Sundelius. Boulder: Westview Press.

———. 1987. Manufacturing and Deindustrialization. In *The New Palgrave: A Dictionary of Economics.* London: Macmillan and New York: Stockton Press.

———. 1989 Third World Competition and De-industrialisation in Advanced Countries. *Cambridge Journal of Economics* 13: 103–120.

Solow, Robert. 1998. Review. *Challenge* 41, no. 1 (January/February): 120–23.

Stiglitz, Joseph. 1997. Reflections on the Natural Rate Hypothesis. *Journal of Economic Perspectives* 11, no. 1 (winter): 3–10.

Thirlwall, A.P. 1992. The Balance of Payments as the Wealth of Nations. Chap. 9 in *The Economics of Wealth Creation.* Aldershot: Edward Elgar.

Thurow, Lester. 1996. The Crusade That's Killing Prosperity. *The American Prospect* (March/April); 54–59.

Tyson, Laura D'Andrea. 1992. *Who's Bashing Whom? Trade Conflict in High-Technology Industries.* Washington, D.C.: Institute for International Economics.

———. 1998. Just How New is The 'New Economy'? *Business Week* (June 1): 30.

Woodall, P. 1996. The Hitchhiker's Guide to Cybernomics. *The Economist,* A survey of the World Economy, September 28.

PART 1

Trade, Macro Policy, and Competitiveness

CHAPTER 2

The Trade Deficit and U.S. Competitiveness

Robert A. Blecker

The U.S. merchandise trade balance soared to $132 billion in 1993, the highest level since the record of $160 billion in 1987.[1] Although the large trade deficits of the late 1980s were widely attributed to an overvalued dollar, the dollar has depreciated substantially since that time. In this context, the persistence of relatively large trade deficits once again raises the question of whether there is an underlying structural decline in competitiveness about which American policymakers should be concerned.

While mainstream economists are divided on whether the U.S. economy has suffered a decline in competitiveness in any sense, they almost unanimously deny that any such decline could be related to the origins and persistence of the trade deficit. Competitiveness cannot matter to the trade balance, we are told, because "the trade deficit is a macroeconomic phenomenon." Based on this premise, economists generally attribute the trade deficit to the fiscal deficit, the (allegedly) low private saving rate, the exchange value of the dollar, mismatched business cycles at home and abroad, the attractiveness of the United States for foreign investment, and virtually any other macroeconomic cause one can think of—anything, that is, but the competitiveness of the nation's industries.

All of these conventional macroeconomic causes of the trade deficit are genuine and important, although the precise degree to which each of them matters can be debated. And, indeed, it is correct that the trade balance is a macroeconomic variable. Where the conventional wisdom has a blind spot, however, is in denying that "structural" factors such as industrial competitiveness can have a macroeconomic impact. This chapter argues that macroeconomic and structural explanations of the trade deficit are not inconsistent with each other, and that our conception of macroeconomics should be broadened to incorporate those structural problems that have some impact at the aggregate level.

When economists say that any explanation of the trade deficit must be macroeconomic, then, they are right, but that does not necessarily mean that we can attribute the causation of the trade deficit entirely to macroeconomic *policies* (especially just U.S. budget deficits) or even to national saving and investment *behavior* more broadly defined. Macroeconomics is simply the study of economic aggregates and economy-wide averages. It includes such factors as productivity growth, labor cost trends, prevalence of trade barriers, preferences for home versus foreign products, and other aspects of what most people would consider to be competitiveness. Structural explanations of the trade deficit are therefore not antithetical to macroeconomic explanations and, indeed, must ultimately find expression in a coherent macroeconomic model.

Indeed, a nation's structural characteristics determine the parameters that influence how any given set of macro policies (both domestic and foreign) impacts on the trade balance. The point, then, is that *structural and macroeconomic explanations of the trade deficit are not mutually exclusive* and must be combined in order to achieve a complete understanding of current trade imbalances in the U.S. economy and elsewhere.

The argument that there is no connection between the trade deficit and competitiveness is often based on the assertion that macroeconomic problems cannot be explained by microeconomic phenomena. Since the trade balance is a macroeconomic variable, it is claimed, it must be explained by macroeconomic factors such as saving and investment rates, fiscal policies, and exchange rates. These macro-level forces supposedly determine a nation's trade balance independently of industry-level performance, which merely influences the composition of the nation's exports and imports.

Such an attitude is curious in a profession in which the reigning theoretical paradigm prides itself on its methodological individualism: the view that everything in the entire economy ultimately reduces to the optimizing decisions of individual agents (households, factor owners, and firms). Surely such a theoretical tradition should admit the possibility that structural changes at the micro-level, if sufficiently widespread, could influence aggregative outcomes to some extent. Stated this way, indeed, the proposition is one that virtually all economists would be forced to accept. The question, then, is why it is so hard for economists to admit that the behavior of the aggregate trade balance (which surely is a macroeconomic variable) may be influenced by the shifting competitive positions of the individual business firms that either sell exports or compete with imports.

Implications of the Equilibrium Method

The reluctance to link trade balances to competitiveness rests fundamentally on the other cornerstone of the neoclassical paradigm: the insistence on using a general equilibrium model in which all markets clear via price adjustment as the benchmark for making welfare judgments and policy recommendations. Among other things, most general equilibrium models assume full employment and balanced trade. The latter assumption is justified by the notion of some sort of automatic adjustment mechanism that eliminates absolute competitive advantages and disadvantages and forces trade to follow comparative advantages. Usually, the story is told in terms of the real exchange rate settling at a value that balances the nation's trade.

In this view, competitive problems may exist (e.g., if a nation's relative rate of productivity growth falls, or if other countries have more rapid qualitative improvement in the quality of their products), but what they affect is *the standard of living at balanced trade* (as determined by the income levels and exchange rates that would prevail in a balanced-trade equilibrium).[2] For example, Hatsopoulos, Krugman, and Summers (1988) argued that the United States has a serious competitiveness problem, defined as the inability of the country to balance its trade while "achieving an acceptable rate of improvement in its standard of living" (299). The logic is that, if a country is not competitive, it can balance its trade only by depreciating its currency or, equivalently, by cutting its workers' real wages. Competitiveness thus influences the real income level (the purchasing power of consumers) that is consistent with balanced trade.

If a country with a trade deficit improves its competitiveness, then the degree of real currency depreciation (or wage cuts) required to balance trade in the long run is reduced, and hence the sacrifice of domestic living standards necessary to restore external balance is diminished. As three influential economists have written,

> The macroeconomic adjustment that the United States faces over the years ahead [in order to reduce the trade deficit] is *linked to the microeconomic issues of competitiveness* in particular products and the general performance of U.S. exports and import-competing industries. How well we compete will determine how far the dollar needs to fall, which in turn makes a major difference to the costs in terms of our standard of living of bringing our trade deficit down. (Dornbusch, Krugman, and Park 1989, 9, italics added)

In this view, micro-level policies are *essential* to alleviate the future costs of adjustment to the U.S. external imbalance. The more the United States can enhance the competitiveness of its own industries, as well as open up markets for its products abroad, the less the dollar and real wages will have to fall in order to eliminate the trade deficit, and the higher will be the sustainable standard of living associated with balanced trade in the long run.

To be sure, this view of competitiveness is not only logical on its own terms, but actually represents a significant concession to those who worry about the problem. Nevertheless, this approach still begs the question of what happens if the adjustments required to achieve and maintain balanced trade do not take place. What determines the trade balance in real historical time, when the actual values of economic variables are not at their hypothetical, general equilibrium levels, and there can be both unemployed resources and imbalanced trade? This question is simply ignored by those who insist on viewing a disequilibrium phenomenon through the lens of an equilibrium theory. But we need to analyze the actual causality involved in determining trade balances in economic systems that are not in a full-employment, balanced-trade equilibrium and that often show little or no persistent tendency in that direction.

Nevertheless, it is important not to lose sight of the fundamental truth that is found in the economists' view of the trade deficit. Since the trade balance *is* a macroeconomic variable, any argument that it is affected by competitiveness must show how competitive problems manifest themselves at the aggregate level. Moreover, any logically consistent theory of the trade balance must respect the famous "national income identity." This dictates that a nation's trade balance must be equal to the difference between what it saves and what it invests at home:[3]

$$\text{Trade Balance} = \text{National Saving} - \text{Domestic Investment.} \qquad (1)$$

"National saving" in turn is the sum of private saving (of households and corporations) plus public sector saving (the government's budget surplus). If the government is running a budget deficit, then public saving is negative, and national saving is less than private saving. By separating the government budget from private saving, we can rewrite (1) as:

$$\begin{aligned} \text{Trade Balance} = {} & \text{Private Saving} - \text{Budget Deficit} \\ & - \text{Domestic Investment.} \end{aligned} \qquad (2)$$

By itself, this identity says nothing about the *direction of causality*—which of the variables (if any) are independent causes of changes in the other variables, or whether all of these variables are responding to some underlying causes that are not incorporated in the identity itself. As Richard N. Cooper has written,

> This relationship is an after the fact identity. It represents an important check on the consistency of any proposed policy, since in order to reduce

> the current account deficit the policy actions must also affect savings and investment in the required way. *But this accounting identity says nothing about the dynamics of the impact of policy actions on the economy.* (1987, 12, italics added)

Cooper's point about logical consistency is well-taken: if one wants to argue that structural changes in a nation's international competitive position can affect its trade balance, one must have a convincing explanation of how such changes affect national saving (relative to domestic investment) so as to satisfy the identity (1) or (2).

Two-Way Causality

At the most general level, all the variables in the national income identity are endogenous variables that are simultaneously determined in the short-run equilibrium of the macroeconomy (which should not be confused with the long-run, full-employment, balanced-trade equilibrium). Exports, imports, personal and business saving, domestic investment, government expenditures, and tax revenue are all functions of various combinations of macroeconomic variables such as income levels, prices, exchange rates, interest rates, and so on. These in turn are influenced by, but not exclusively determined by, government fiscal and monetary policies. Even the level of the government deficit is not determined by tax and spending policies alone, but also depends on the levels of income and employment (which influence tax receipts and entitlement expenditures), as well as the monetary policies of the Federal Reserve Board, which determine the interest rates at which the government can finance its debt (not a trivial consideration in a country with a $4 trillion debt!).

What all of this means is that causality essentially runs both ways in the macro identity. Changes in domestic saving and investment behavior and fiscal policies do, of course, affect the trade balance. These factors operate partly through the level of national income, which strongly influences import demand, and partly through net foreign capital flows and exchange-rate fluctuations. But changes in the competitiveness of a nation's tradeable goods also affect national saving and investment as well as exports and imports. More competitive nations earn higher incomes and employ more people, ceteris paribus, than less competitive nations, and these higher levels of income and employment translate into greater private saving and tax revenue. Competitiveness matters precisely because it affects both sides of the macro identity, not just the trade balance side.

Suppose, for example, that a country's trade performance improves as a result of innovations that give domestic products a greater competitive advantage in either quality or cost, or from the opening up of a new foreign market. The immediate effect will be to increase the country's exports or reduce its imports, and thus to improve the trade balance, for any given exchange rate and level of national income. Let us trace out exactly how this will affect national saving and domestic investment.

Assuming that the country has some excess industrial capacity and less-than-full employment, as is normally the case, higher net export demand increases domestic output and employment. As output and employment rise, national income grows. The increased national income will be divided between higher capital and labor incomes; in general, both will increase to some extent. Now these higher incomes in turn result in higher private saving, both corporate and personal. Since corporate saving accounts form the lion's share of gross private saving in the United States, the increase in corporate profits is especially important for total saving to rise.

In addition, higher private incomes imply higher tax revenue for the government, since most taxes are proportional to either income or spending (and spending tends to rise with income). Since government expenditures are unlikely to rise when private incomes increase (and are likely to fall if higher employment lessens social welfare costs), the government budget surplus should increase (or the deficit decrease). Thus both parts of national saving, private and public, should benefit from greater international competitiveness.

It is also likely that business investment will be stimulated to some extent by increased consumer demand and higher capacity utilization in industry, as well as by increased corporate profitability (which provides incentives to invest and relieves potential financing constraints). Since increased domestic investment will absorb some part of the increased national saving, the net improvement in the saving-investment balance will be diminished to the extent that investment also rises. However, in a stable macroeconomic system,[4] the induced increase in domestic investment will normally be less than the induced increase in national saving (private saving plus the government surplus). This implies that the right-hand side of identity (1) should increase to some extent as national income adjusts to its new, higher level.

At the same time, some part of the increased national income will be spent on imports. This leakage to imports reduces but does not eliminate the positive gain in the trade balance from improved competitiveness (or access to foreign markets), since only a fraction of every dollar of increased income will be spent on imports. In addition, the greater the competitiveness of domestic products, the smaller this fraction should be.[5] Thus, an improvement in competitiveness should be expected to raise the equilibrium trade balance *and* national saving relative to domestic investment—thus increasing *both* sides of identity (1)—although the net increase in the trade balance will be somewhat less than the hypothesized initial increase in exports (or decrease in imports).

In the case of a deterioration in competitiveness, this mechanism will work in reverse. Domestic output and national income will be reduced, employment will fall, and both corporate profits and workers' wages will be lower. Both private saving and government tax revenue should then be expected to fall, thus reducing national saving (by relatively more than investment falls). The trade balance will worsen, although this effect will be partially mitigated by reduced demand for imports due to lower national income. The government budget surplus will also tend to fall (the deficit to rise), creating an apparent twin relationship between the trade and budget deficits—but without a change in fiscal policy being responsible.

Many analyses of trade imbalances implicitly assume full employment. For example, this is evidently assumed by McCulloch and Richardson (1986, 51) when they assert that improved competitiveness would mean "higher-productivity jobs" but "not necessarily more jobs." If one assumes full employment, then of course "more jobs" are ruled out. But in spite of the tendency among economists to assume that there is always full employment in the "long run," actual advanced capitalist economies tend to oscillate between periods of relatively high and relatively low employment, rather than to stay at full employment unless disturbed. In this case, better trade performance can help provide "more jobs" as well as "better jobs."

Exchange Rate Adjustment and Living Standards

In theory, the positive effects of improved competitiveness just described could be completely offset by a sufficient appreciation of the nation's currency. Indeed, this is exactly what is argued by those who claim that competitiveness does not matter to the trade balance:

> The ingredients of industrial competitiveness cannot be regarded as *independent* determinants of trade performance. In particular, induced movements in exchange rates tend to offset the aggregate consequences of any improvement in industrial competitiveness. Unless national saving rises relative to the demand for funds for private investment, any increase in the competitiveness of an industry will be accommodated mainly through a *further appreciation* of the dollar. (McCulloch 1986, 26, italics in original)

The qualifying phrase "Unless national saving rises" in the last sentence is crucial; as explained above, this is exactly how an increase in competitiveness (or opening up of foreign markets) can improve the trade balance.

While McCulloch implies that exchange rate movements immediately offset changes in competitiveness, Robert Z. Lawrence takes the more defensible position that such adjustments only occur in the long run:

> Changes in the relative trade performance of American industries will only affect the trade balance in the short run. The trade deficit may rise temporarily, but the loss of foreign markets will put downward pressure on U.S. wages and prices, and, more important, will tend to depress the exchange value of the dollar . . . to the point where the trade deficit turns around and moves back to an equilibrium determined by the country's fundamental spending-saving behavior. (1989, 29)

In spite of his efforts to emphasize the long-run adjustment mechanism, however, Lawrence effectively concedes that the trade balance *is* affected by industrial competitiveness in the short run—defined as the period in which exchange rates have not sufficiently adjusted. Once the problem is posed in this way, however, it becomes relevant to ask: how long is the "short run"? How long will it take for the exchange rate to settle at its new, long-run equilibrium level, following a change in competitiveness?

In order for the exchange rate to adjust to a level that would offset competitive advantages (or disadvantages), it is necessary for exchange rate changes to be driven by the need to restore balanced trade. That is, when a country's trade balance improves, its currency would have to appreciate; when a country's trade balance worsens, its currency would have to depreciate. Before flexible exchange rates were adopted in the early 1970s, this is generally how economists anticipated that a system of flexible rates would work. But the reality has turned out to be quite different. Flexible exchange rates are determined by the conditions for *overall* balance-of-payments equilibrium, not balanced merchandise trade or even a balanced current account (which includes trade in services, net interest payments, and unilateral transfers). This means that capital account transactions—that is, international investment, borrowing, and lending—play an important role in determining currency fluctuations.

In fact, international capital flows have turned out to be far *more* important than current account transactions (trade in goods and services) for determining exchange rates. The sheer volume of international financial flows dwarfs the value of trade flows. According to Levich (1988, 220), foreign exchange transactions in the three largest trading centers (London, New York, and Tokyo) alone totaled $188 billion *per day* as of March 1986, and "worldwide foreign exchange could possibly exceed $250 billion per day or more than $60 trillion per year." In contrast, the total value of U.S. merchandise trade (exports plus imports) in 1986 was $592 billion for the entire year, or less than $2 billion per day.

It is now generally recognized that the main determinants of fluctuations in currency values, at least in the short run, are located in international financial markets, not in international goods markets. This insight led economists in the 1970s to develop theories in which underlying macroeconomic "fundamentals" (such as money supplies, budget deficits, interest rates, and inflation rates) explain changes in exchange rates. By the end of the 1980s, however, international economists had to admit that even these theories could not explain much of the volatility of exchange rates over the preceding two decades. In particular, a number of economists (e.g., Dornbusch 1988, 1989; Frankel 1990; Krugman 1989a) have concluded that the sharp rise in the U.S. dollar in the mid-1980s went far beyond anything that could be explained by such fundamentals.

The assertion that, in the long run, the value of the dollar must adjust to offset any change in real competitive advantages seems to be based on outmoded a priori reasoning rather than any evidence on the actual behavior of exchange rates in recent years. Certainly the experience of the 1980s suggests that the dollar can stay above a level that would be consistent with balanced trade for at least a decade. But if exchange rates do not do their job, then the "short run" can be quite long in practice. If the dollar does not adjust to its hypothetical "long-run" equilibrium value faster than the underlying competitive conditions themselves change, then actual trade deficits can be continuously affected by those conditions for an indefinite period of time.

Moreover, even if the nation's currency does eventually adjust to its new equilibrium value, this new value will reflect the changed competitive conditions. This point is crucial, because the value of the nation's currency is an important determinant of its cost of living and therefore of the purchasing power of domestic incomes. To the extent that people consume imported products, or products that are manufactured with imported inputs, a higher value of the currency means a higher standard of living for domestic residents; a lower value means the opposite. Therefore, even if one is optimistic about the effectiveness of exchange rate adjustment, one should still be concerned about a nation's underlying competitiveness.

The fact that it can take a long time for the exchange rate adjustment process to work is especially significant because of its dynamic implications. *By the time the exchange rate eventually adjusts, a country is not starting from the same initial conditions as when the competitive improvement (or deteriora-*

tion) originally occurred. In the intervening years, the more competitive nation will have invested in more new capital and R&D, increased its labor productivity faster, and raised its technological advantages further, while the less competitive nation will have fallen behind in all of these respects. The eventual appreciation or depreciation of the currency will not necessarily be sufficient to reverse these dynamic gains or losses and therefore will not generally return a country's economy to the position it was in before the change in competitiveness.

The Fiscal and Trade Deficits: Twins or Cousins?

In the 1980s, it became popular to argue that the U.S. trade deficit could not result from competitive problems because it was merely a result of the federal government budget deficit. Using the macro identity, it was argued that the fiscal deficit caused a decrease in national saving, which in turn caused a nearly equal decline in the trade balance. This view was promulgated by many prominent economists, such as former Reagan economic adviser Martin Feldstein. It also appeared to be supported by some theoretical models, such as the Mundell-Fleming analysis[6] in which a fiscal deficit crowds out an equal amount of net exports in a small country with perfect capital mobility and a flexible exchange rate.

While this so-called twin deficit view found its way into newspaper editorials and pundits' columns, in fact it always rested on shaky empirical grounds: not that there was no connection between Reagan's budget deficit and the trade deficit in the early 1980s, but the relationship was far less than the $1 for $1 implied by the twin deficit terminology. A number of studies attempted to quantify the twin deficit relationship by estimating how much the increased budget deficit contributed to the increased trade deficit in the early 1980s. While some of these were advertised as supporting the twin deficits view, in fact most of the studies found at most only a partial causal connection between the rise in the U.S. federal government budget deficit and the increased U.S. trade deficit in the early 1980s.

A systematic comparison of international macroeconomic models conducted by Bryant et al. (1988) revealed that different models constructed according to different assumptions yielded very different quantitative (and sometimes even qualitative) results about the effects of shifts in fiscal policies. For example, while most of the models in the study agreed that a 1 percent of GNP fiscal stimulus in the United States would lower the current account balance, estimates of the magnitude of this effect (cumulated over six years) ranged from $3 billion to $53 billion.

The model that generated the *largest* estimated effect of fiscal expansion on the trade deficit in the Bryant et al. study was the Multicountry Model (MCM) of the Federal Reserve Board. Helkie and Hooper (1988, 48) applied this model to analyze the underlying macroeconomic causes of the increased U.S. current account deficit from 1980 to 1986.[7] They estimated that U.S. fiscal expansion accounted for about a $70 billion rise in the current account deficit, or just under half of the actual increase of $143 billion, holding domestic monetary and foreign fiscal policies constant.[8] Thus a *high-end* estimate of the twin deficit relationship is that *less than half* of the 1980 to 1987 rise in the trade deficit can be explained by the rise in the U.S. budget deficit! And most of the other macro models surveyed by Bryant et al. (1988) yielded substantially *smaller* estimates of this relationship. As Hooper concluded elsewhere,

> Fiscal policy, or the twin deficit relationship, is a significant part of the story, but evidently explains *no more than half* of the decline in our net foreign saving [i.e., current account balance]. The remaining half must be explained by fundamentals other than macroeconomic policy that led to a decline in private domestic saving relative to investment. (1989, 38; italics added)

In another study of international imbalances, Sachs (1988, 645) found that "a sustained, bond-financed U.S. fiscal expansion (an increase in federal spending on goods and services [of 1 percent of gross national product]) . . . worsens the U.S. trade [current account] balance . . . by an average of 0.31 percent of GNP over three years." Although Sachs's result is obtained from simulations of a large-scale general equilibrium model of the world economy (the McKibbin-Sachs Global Model—see McKibbin and Sachs 1989), an almost identical result is obtained in a much simpler econometric exercise by Bernheim (1988). Using a single-equation model in which only the budget deficit and the GNP growth rate (current and lagged) are used to explain the current account, Bernheim finds that "a $1 increase in government budget deficits leads to roughly a $0.30 rise in the current account deficit" (2) in the United States. Sachs's and Bernheim's results imply that a $100 billion reduction in the budget deficit would eliminate only about $30 billion of the current account deficit—even assuming an offsetting monetary expansion, as Sachs does. On this basis, Sachs concluded

> The U.S. fiscal expansion was only *one* of the reasons for the widening of the U.S. current account deficit. Completely *eliminating* the U.S. budget deficit, other things being equal, would remove no more than half the

> current external gap. . . . Balancing the U.S. current account will therefore require policy actions or other economic events . . . beyond balancing the U.S. budget. (Sachs 1988, 646–47; italics in original)

Curiously, even some economists who claim to have evidence for the twin deficits hypothesis in fact have evidence against it. For example, the study by Rosensweig and Tallman (1993) found that about 30 to 40 percent of the variation in the U.S. trade balance in the 1980s could be explained by variations in the fiscal deficit. Although the authors assert that their evidence is "consistent with . . . the typical twin deficits explanation" (590), it requires an incredible stretch of the English language to label a 30 to 40 percent connection "twins." To be sure, the fiscal and trade deficits were probably at least cousins in the mid-1980s, when both increased notably. This is the period emphasized in the studies just cited. But taking a longer historical view, these two deficits have not always been as correlated as they seemed at that time. In the next section, we examine what the longer term record shows.

Evidence for a Secular Decline in Competitiveness

A Retrospective on the 1980s

Table 2.1 gives some perspective on the changes in the U.S. trade balance and its macroeconomic determinants in the 1980 through 1990 period, ending before the trough of the recent recession in 1991. Data are given for 1980 and 1990, and for one intermediate year, which is either 1985 (for national income and exchange-rate data) or 1987 (for trade balance data). These different break points are chosen because of the different timing of the trends in U.S. and foreign growth, in the value of the U.S. dollar, and in the trade balance. In particular, the trade balance seems to respond with a lag especially to changes in relative prices.

From 1980 to 1987, the trade balance fell by almost 4 percent of GNP by any of the three measures shown (a bit less for the total merchandise balance in current dollars, which was aided by the fall in prices of oil imports in the mid-1980s). Undoubtedly, the primary direct cause of this deterioration was the enormous rise in the real value of the dollar between 1980 and 1985,[9] which averaged about 56 percent by the Federal Reserve Board index (which emphasizes the G-10 industrial countries and uses consumer price indexes [CPIs] to adjust for inflation) and about 35 to 38 percent by the other indexes shown. However, sluggish growth in foreign countries—especially those that

TABLE 2.1. U.S. Trade Balance and Related Macroeconomic Indicators, 1980, 1985 or 1987, and 1990 (index numbers, 1980 = 100, except as noted)

	1980	1987	1990
Trade balance (percentage of GNP)			
Merchandise[a]	−0.8	−3.5	−1.9
Nonoil, nonagricultural[a]	0.5	−3.2	−1.5
Current account[b]	0.5	−3.4	−1.6
	1980	**1985**	**1990**
National Incomes			
United States[c]	100	115	133
Other industrial countries[c,d]	100	112	133
Western Europe	100	108	126
Japan	100	121	153
Developing countries[e]	100	114	135
Latin America[e]	100	102	112
Real exchange rate (value of dollar)			
FRB G-10 index[f,g]	100	156	102
Dallas Fed 101 countries index[g,h]	100	135	110
IMF wholesale price index[i]	100	136	94
IMF unit labor cost index[i]	100	139	77

Source: Trade balance data are taken from the U.S. Department of Commerce 1986; and updates in the *Survey of Current Business*, various issues. Other sources are as noted.

[a]Trade balances measured in current dollars on a national income and product account basis.

[b]Measured by net foreign investment in the national income and product accounts.

[c]Measured by gross domestic products (GDP) in constant 1985 dollars. From OECD 1991, 172.

[d]Total OECD minus U.S. and Turkey. Includes some countries not listed separately (Canada, Australia, New Zealand).

[e]From index in IMF, *International Financial Statistics, 1991 Yearbook*, 163. Indexes extrapolated to 1990 at average annual rates for 1980–89. Latin America is defined as all developing countries in the Western Hemisphere.

[f]From U.S. Council of Economic Advisors 1991, 410, based on data from Board of Governors of the Federal Reserve System.

[g]Nominal exchange rates are adjusted by consumer price indexes.

[h]Unpublished data from Federal Reserve Bank of Dallas.

[i]From IMF, *International Financial Statistics, 1991 Yearbook*, 110–11. The normalized unit labor cost index was used.

traditionally bought the largest volumes of U.S. exports (Western Europe and Latin America)—also contributed substantially to the increased trade deficit in the early 1980s.

By 1990, these proximate causes of the large trade deficits had been largely reversed or eliminated. In general, foreign income growth speeded up (especially in Western Europe, but also to some extent in Japan and the developing countries). The gross domestic product (GDP) of all other industrialized countries in 1990 was 33 percent above its 1980 level—exactly the same as for U.S. GDP. Latin America, depressed by the "stabilization" and "structural adjustment" policies adopted in response to the debt crisis, was the only major region (in terms of U.S. export markets) that lagged far behind U.S. growth over the whole decade (although Latin America has recovered considerably since 1990).

At the same time, the dollar fell just about as far in real terms in the late 1980s as it had risen in the early 1980s. The Federal Reserve Board's CPI-adjusted G-10 index shows the dollar back to a mere 2 percent above its 1980 level in 1990, while the broader Dallas Fed index (which includes many developing country currencies and is also CPI-adjusted) shows the dollar about 10 percent higher in 1990. Both of the International Monetary Fund (IMF) indices show the dollar lower in 1990 compared with 1980, especially when the nominal exchange rates are adjusted by normalized unit labor costs rather than by wholesale prices.

In spite of these reversals, the trade deficit remained about 2 percent of GNP worse in 1990 than it had been in 1980 by two of the measures shown (nonoil import, nonagricultural export merchandise trade—which is mostly trade in manufactures—and the current account) and about 1 percent worse for the total merchandise balance (which was helped by the declining trend in oil import prices over the whole decade). This suggests a notable structural deterioration in the U.S. trade position, centered in trade in manufactures.

Table 2.2 shows the trends in U.S. real wages and unit labor costs relative to major competitor nations from 1975 to 1990. Throughout this period, real hourly compensation in manufacturing was stagnant or falling slightly in the United States, while real hourly compensation in the other nations shown[10] continued to rise steadily. The fact that U.S. real hourly compensation declined absolutely without achieving balanced trade suggests a particularly grave competitiveness problem by the Hatsopoulos-Krugman-Summers definition. The United States has not been able to balance its trade in spite of *substantially reducing its standard of living* in terms of hourly labor compensation in manufacturing.[11]

U.S. (nominal) unit labor costs in manufacturing rose slightly relative to foreign unit labor costs from 1975 to 1980 in own currency terms, but fell relatively in dollar terms due to the depreciation of the dollar in the late 1970s. In the 1980s, U.S. unit labor costs fell relative to foreign costs throughout the decade in own currency terms. U.S. unit labor costs rose relative to foreign unit labor costs in dollars from 1980 to 1985 while the dollar was rising, but by 1990, U.S. unit labor costs had fallen relative to those of the other countries in dollar terms beyond where they were in 1980. The fact that the U.S. trade balance remained in deficit by nearly 2 percent of GNP in spite of *falling* U.S. relative labor costs suggests an especially acute qualitative competitiveness problem.

TABLE 2.2. U.S. Real Hourly Compensation and Unit Labor Costs in Manufacturing Compared with Major Competitor Nations (indexes, 1975 = 100)

	1975	1980	1985	1990
Real hourly compensation				
United States	100	101	102	96
Canada	100	109	115	117
Japan	100	101	107	122
Germany (West)	100	118	122	140
France	100	118	134	142
United Kingdom	100	109	116	127
Italy	100	105	111	117
Korea (South)	100	167	222	428
Nominal unit labor costs (U.S. relative to other countries)				
In own currencies				
Industrial countries[a]	100	103	98	87
Industrial countries[a] plus Korea and Taiwan	100	101	95	84
In U.S. dollars				
Industrial countries[a]	100	95	128	75
Industrial countries[a] plus Korea and Taiwan	100	94	124	73

Source: Hourly compensation in national currency is from U.S. Department of Labor 1991. Compensation in national currency was deflated by consumer price indexes from IMF, *International Financial Statistics,* 1991 Yearbook and October 1991. Relative unit labor cost indexes are taken from unpublished BLS data on "Trade Weighted Output per Hour, Hourly Compensation, and Unit Labor Costs in Manufacturing, 1970–90" (1991).

[a]Industrial countries included are Canada, Japan, Denmark, France, (West) Germany, Italy, Netherlands, Norway, Sweden, and United Kingdom.

The Falling Trend of the Dollar

Another perspective on U.S. competitive decline can be gained by comparing the trends in the trade balance with the trends in the exchange rate. Since the collapse of the Bretton Woods system of fixed exchange rates in 1973, the dollar had to depreciate in real terms in order to prevent growing trade deficits—and those deficits have grown precisely whenever the dollar did not depreciate or did not depreciate sufficiently.

Figure 2.1 shows the merchandise trade balance and the current account balance, both measured as percentages of GDP.[12] Note that the current account remained balanced throughout most of the 1970s, in spite of some fluctuations. The merchandise trade balance worsened notably in the late 1970s, but this was mainly due to the rising price of oil imports (the nonpetroleum trade balance, not shown in figure 2.1, did not have a downward trend in the 1970s). The current account balance was higher than the merchandise balance as a result of a trade surplus in services, including positive net investment income from abroad. In the 1980s, however, both balances

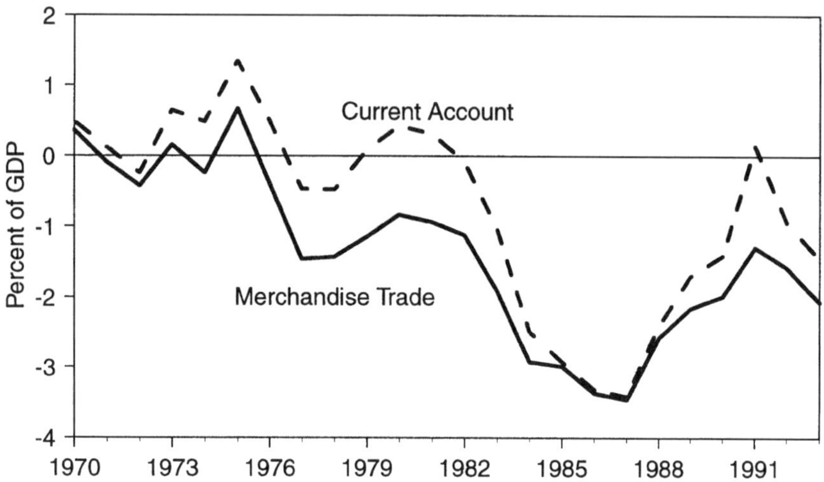

Fig. 2.1. Alternative measures of the U.S. trade balance, 1970–93. (Author's calculations from U.S. Department of Commerce, Bureau of Economic Analysis, National Income and Product Accounts [NIPAs]. The current account is measured by the NIPA concept of net foreign investment.)

turned sharply negative together, and the gap between them shrank as net investment income from abroad dried up (due to increasing U.S. foreign debt). By both measures, the trade deficit peaked in 1987.

In the late 1980s and early 1990s, both measures of the trade balance showed improvement as a result of two factors: the slowdown in U.S. economic growth (culminating in the 1990–91 recession), and the post-1985 depreciation of the U.S. dollar (see Blecker 1991b). However, the apparent current account surplus in 1991 was the spurious result of transfer payments received from foreign nations (especially Germany, Japan, and Saudi Arabia) to pay for Operation Desert Storm and did not reflect any permanent improvement in the U.S. trade position. Once the U.S. economy recovered from the recession (and the transfer payments ceased), the trade balance worsened again by both measures in 1992–93, with the merchandise trade deficit again reaching about 2 percent of GDP.

Figure 2.2 shows two indices of the real value of the dollar (these are two of the indices also shown in Table 2.1). The first is the Federal Reserve Board index of the trade-weighted value of the dollar against the G-10 industrial country currencies, corrected for consumer price inflation. The second index is the IMF real effective exchange rate index covering 16 industrial countries based on relative normalized unit labor costs. Both measures show the same

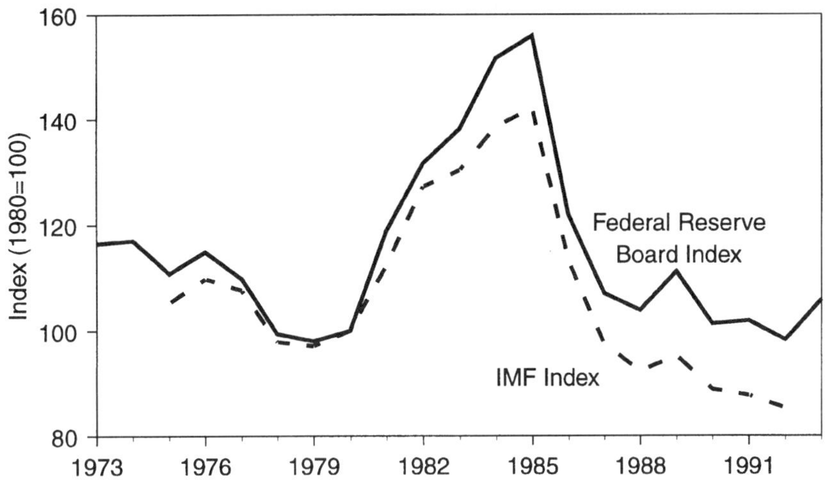

Fig. 2.2. Alternative measures of the real value of the U.S. dollar, 1973–93. (Author's calculations from U.S. Council of Economic Advisors, *Economic Report of the President,* February 1994; and IMF, *International Financial Statistics,* 1994 Yearbook.)

strong rise in the dollar in 1981 through 1985 and the same sharp decline thereafter, but the IMF index shows a more pronounced decline in the early 1990s (due to the fall in U.S. relative wages discussed earlier) while the Fed index shows the dollar stabilizing at that time.

Comparing figure 2.2 with figure 2.1, we see that the dollar was more or less continuously falling while the current account remained essentially balanced (or in surplus) throughout most of the 1970s. To be sure, there were current account deficits in 1971 and again in 1977–78, but they were relatively small and more than offset by surpluses in the rest of the decade—thanks to the continued depreciation of the dollar. The fact that a continuous real depreciation of the dollar was necessary to maintain current account balance and to keep the merchandise trade deficit from worsening indicates an *underlying declining trend* of U.S. competitiveness in the 1970s.

The problem in the early 1980s was not simply that the dollar rose, but more precisely that *it rose when it needed to fall further.* To make this point, figure 2.3 extrapolates the 1973–79 trend of the Fed index into the 1980–93 period, assuming a constant proportional rate of change.[13] Along this trend line, the dollar fell at an average rate of about 3 percent per year from 1971 to 1980. Extending this trend line to 1993 gives a rough indication of the exchange rate adjustment that would have been required to maintain current account balance in the 1980s.

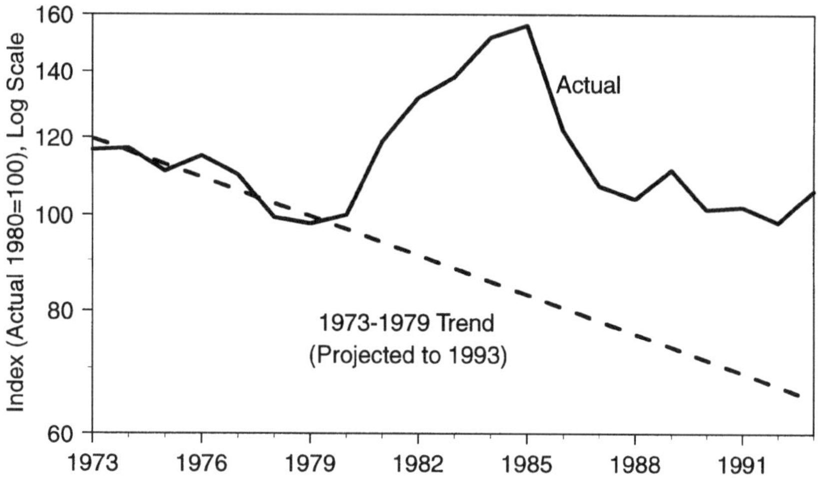

Fig. 2.3. Real value of the U.S. dollar, actual 1979–93 versus trend for 1973–79 projected in 1993. (Author's calculations from U.S. Council of Economic Advisors, *Economic Report of the President,* February 1994.)

Viewed against this benchmark, the rise in the dollar in the mid-1980s is truly astounding. From 1980 through 1985, the dollar moved in the *opposite* direction from that which would have been necessary to offset declining competitiveness. Moreover, this way of viewing the problem shows that relying on dollar depreciation to solve the trade deficit is like shooting at a moving target. Although the dollar fell to about its 1980 level in the early 1990s, it was still far above the extrapolated 1970s trend line; the dollar would have to fall by about 40 percent in real terms in order to return to that trend.

It is also important to realize that these exchange rate indices, which emphasize industrial country currencies, do not give a complete picture of the value of the dollar. This is because a rising share of U.S. imports and of the U.S. trade deficit is accounted for by the developing nations, especially the newly industrializing countries (NICs) of East Asia and Latin American countries like Mexico. On the whole, the dollar has not depreciated as much against these countries' currencies since 1985 as it has against the industrial country currencies, and that difference may account for the shift in the composition of the U.S. trade deficit toward the NICs as discussed below.

Income Elasticities and Time Trends

Economists have developed more rigorous methods for detecting the presence of structural competitive problems. These methods involve the statistical estimation of what are called import and export demand functions. These are mathematical expressions that show how a country's imports and exports depend on other variables such as the relative prices of home and foreign products (which are heavily influenced by exchange rates) and the income that domestic or foreign residents have to spend on imports or exports, respectively. Other variables that are thought to explain the quantities of imports and exports can also be included in these demand functions, if desired.

The use of import and export functions yields estimates of the income elasticities of U.S. import and export demand. These elasticities are the percentages by which the quantity of our imports or exports rises for a 1 percent rise in domestic or foreign income, respectively. If the income elasticity is higher for imports than for exports, this implies that we can only maintain balanced trade by either (a) continuously depreciating the dollar in order to keep our demand for imports from growing too fast (and to make our exports grow faster) or (b) keeping our income growth rate lower than the rest of the world's, by about the same proportion as the ratio of our income elasticities for exports and imports, if the dollar is not allowed to fall (Thirlwall 1979).

Table 2.3 gives the estimates for the income elasticities of imports and exports from a variety of recent studies, including ones done expressly for this chapter. Most of the estimates show a notably higher income-elasticity of U.S. demand for imports compared with foreign demand for U.S. exports, thus confirming the existence of a competitiveness problem in the U.S. case. Of the two studies that do not find a large difference in income elasticities, at least one (Helkie and Hooper 1988) includes another variable that picks up the effects of structural competitive decline (it uses the ratio of U.S. to foreign capital stocks to represent supply side conditions). The other exception

TABLE 2.3. Estimates of Income Elasticities of U.S. Export and Import Demand

Author(s)	Exports	Imports
Krugman and Baldwin (1987)	2.42	2.87
Helkie and Hooper (1988)[a]	2.19	2.11
Cline (1989)	1.70	2.44
Lawrence (1990)[b]	1.60	2.47
Blecker (1992b)	1.67	2.68
New estimates for this chapter[b]	1.38	2.22

Source: See bibliography for references to older studies, and the appendix in this chapter for explanation of new estimates.

[a]Model includes another variable (the ratio of U.S. to foreign capital stocks) that picks up structural trends in competitiveness.

[b]Model includes an adjustment for computer price measurement.

(Krugman and Baldwin 1987) is the oldest of the studies in the table and does not include data from the late 1980s.

Of the studies that do find a difference in the income elasticities, the work of Cline (1989) is especially noteworthy because it is based on a disaggregated study of U.S. trade with several other specific countries and regions. Also, Lawrence's (1990) results are important because he corrected for biases in the measurement of the "real" (inflation-adjusted) quantity of computer imports and the prices of computers. These biases are thought to exaggerate the real volume of U.S. imports to some extent.[14] For this reason, the new estimates prepared for this chapter, which are given in table 2.3, are done with a computer correction as well.[15]

Another econometric approach involves the estimation of time trends in the import- or export-demand function. That is, it is possible to test whether imports (or exports) tend to rise (or fall) over time by a greater amount than can be explained by the observable variables such as relative prices (including exchange-rate effects) and relative incomes. According to the results in Blecker (1992a, 1992b), there is fairly robust evidence for a positive time trend in U.S. demand for nonpetroleum imports. The real volume of U.S. imports (in inflation-adjusted terms) tends to grow by about 2 percent per year more than can be explained by the observable problems. Statistical problems, however, preclude any firm conclusions about a time trend for U.S. exports. These findings suggest that the greatest U.S. competitive problems lie on the import side, where qualitative improvements in foreign productive capabilities have caused a structural shift toward imports in U.S. consumers' and firms' purchasing patterns.

Income Adjustment and Absolute Advantages

What are the implications of the declining tendency of U.S. competitiveness identified in the previous section? One approach, discussed earlier, is to focus on the implied need for the dollar to depreciate in real terms in order to maintain balanced trade in the long run. If these hypothesized long-run adjustments do not take place, how declining competitiveness manifests itself in actuality remains to be analyzed. In this vein, another approach emphasizes the adjustment of national income levels and growth rates assuming that real exchange-rate adjustment does not take place. This alternative approach was pioneered by British Keynesian economist A. P. Thirlwall (1979) in his theory of balance-of-payments constrained growth. The income-adjustment view has been extended recently by Dosi, Pavitt, and Soete (1990) in their work on "technological gaps" and "absolute competitive advantages." What follows is a brief effort to explain this alternative view of the international adjustment process.

Assuming that countries cannot maintain trade deficits (or surpluses) indefinitely, and that exchange rate adjustments do not necessarily operate, then relative income levels (or growth rates) are constrained in the long run by the requirements of balancing trade. The crucial constraints are the income elasticities of demand for a country's exports and imports, as discussed in the preceding section. If a country like the United States has an income elasticity of demand for its imports that is higher than its trading partners' income elasticity of demand for its exports, then the former country must grow at a slower rate in order to prevent growing trade deficits with its trading partners. While macro policies can foster an "adjustment" to such a problem, they cannot overcome it. On the contrary, macro policy responses designed to cure trade deficits—such as contractionary fiscal policies—are likely to just reinforce the tendency for slower growth in the less competitive country. Only direct efforts to address problems of relative costs and quality, as well as market access barriers, can overcome this kind of constraint.

The concept of a trade balance constraint on national income implies that the most important question for the United States is not whether it can balance its trade, but *how* to balance its trade. One way to balance trade is to make domestic incomes and expenditures grow sufficiently slowly in order to hold down the growth of imports from the rest of the world. This is the belt-tightening or austerity path, and it will cut domestic living standards just as surely as a continuous real depreciation of the dollar (perhaps even more surely). Yet this is exactly what a reduction in the government budget deficit accomplishes, unless accompanied by other measures to stimulate demand. The alternative is for the United States to adopt industrial and trade policies that can relieve the trade balance constraint, combined with expansionary foreign macro policies and market-opening measures to increase demand for U.S. exports.

In *The Economics of Technical Change and International Trade*, Dosi et al. (1990) concluded that much international trade in manufactures is determined by the technological innovativeness of the various national firms rather than by comparative cost factors. Technological innovation includes the development of new and improved products as well as more efficient production processes (which lower costs). Countries with superior technologies in sectors such as computers, aircraft, chemicals, machinery, and scientific or medical instruments can export these goods in spite of high wages. The income-adjustment view says that such absolute technological advantages determine whether a country will have a tendency to run overall trade surpluses or deficits, and thus whether it will have a relatively loose or tight trade balance constraint on its income level and growth rate. The wider the range of sectors in which a country has such absolute superiority, the higher are the real wages, per capita income, and growth rate that are consistent with balanced trade.

Even in this new view, adjustments of relative wages and exchange rates can still permit much trade to take place according to comparative advantages. Thus, for example, the United States will be a net importer of goods like textiles and steel in which it has lost comparative advantages (although the more productive and high-end segments of these domestic industries will remain viable). These same adjustments also allow countries with superior technologies to have higher real wages and living standards. It is precisely because their wages are higher that these countries must be net importers of goods in which their absolute productivity advantage is relatively low (i.e., in which they have a comparative disadvantage). To this extent, the new view accepts part of the exchange rate adjustment view discussed earlier. But the new view adds that there is a second type of adjustment process, which is the adjustment of relative national incomes. A country that has absolute competitive advantages in a wide range of manufacturing industries will be able to grow faster with balanced trade, while a country with absolute competitive disadvantages will have to grow more slowly in order to have balanced trade.[16] Furthermore, there are positive feedbacks from rapid growth and negative feedbacks from slow growth, which tend to reinforce the initial gaps in absolute competitiveness. For example, a rapidly growing country will generally be able to sustain higher rates of saving and investment; it will thus be able to upgrade its capital equipment and raise its productivity faster. A slowly growing country will generally have low saving and investment rates and will therefore be saddled with an aging capital stock and sluggish productivity growth. These self-reinforcing mechanisms in international competitiveness are examples of what is called *cumulative causation*—which essentially implies that it is harder for a country to catch up once it has fallen behind.

Alternative Interpretations of Declining Competitiveness

Factor Costs

The declining trends in U.S. competitiveness suggests an agenda for research to determine their sources. Judging from the data on relative wage changes and unit labor costs in table 2.2, it does not appear that the United States has lost out in comparative labor cost terms, at least compared with other industrialized countries. If anything, the United States has gained ground in this respect—at the cost of reducing its workers' ability to improve their standard of living without working more jobs or longer hours. Some authors put the blame on high capital costs, not so much for their direct effect on overall costs, but rather for their allegedly large, negative impact on investment and thus on productivity growth and qualitative improvement (see, e.g., Hatsopoulos, Krugman, and Summers 1988). However, this argument requires some extreme assumptions about the importance of the cost of capital versus other factors for explaining investment, assumptions that are not robust in empirical studies that separate cost of capital factors from accelerator, profitability, and internal financial variables (see, e.g., Fazzari and Athey 1987, Fazzari, Hubbard, and Petersen 1988; and Fazzari 1993).

Moreover, the causes of high capital costs in the United States in the 1980s are subject to dispute. While the traditional view ascribes higher capital costs to a low saving rate and tax disincentives, the accuracy with which the U.S. saving rate is measured has been questioned by Eisner (1991), Block (1990), and Blecker (1990, 1991a), among others. It has also been argued (McCauley and Zimmer 1989) that the higher cost of capital in the United States is explained by differences in financial institutions and a larger risk premium on investment in the United States compared with Japan or Germany. In addition, new theories of investment (summarized in Dixit 1992) imply that investment should be less sensitive to the cost of capital and more sensitive to the stability of the economic environment when entrepreneurs are uncertain about future returns to current investment projects and when those projects involve sunk costs—problems that are ignored in the traditional view of investment.

A Product-Cycle View

An alternative hypothesis about declining U.S. competitiveness would focus first and foremost on changes in the global pattern of technological innovation, development, and diffusion. As Krugman (1979) showed in his formalization of the product-cycle model, a country like the United States that specializes in technologically innovative, manufactured products has its equilibrium wage level determined by the difference between the rate of new product innovation and the rate of technology transfer to less developed countries. Furthermore, in a world with more than one innovator nation, each innovator gets more rents from its new products when it has a greater monopoly over new innovations.

In recent decades, the United States position has eroded on every front. First, the U.S. has clearly lost its unique dominance in the field of technological innovation. Japan in particular, but Germany and some other European countries as well, have been catching up with or surpassing the United States by various measures of technological leadership. For example, the U.S. share of total business sector R&D spending among the industrialized countries has fallen from about two-thirds in 1967 to barely over one-half in 1987, while the Japanese share soared from less than one-tenth to over one-quarter (Dosi, Pavitt, and Soete 1990, 45). And the share of GNP devoted to nondefense R&D in the United States is only about two-thirds of the same share in Japan or (preunification) West Germany (U.S. National Science Foundation 1990).

Furthermore, as Wright (1990) points out, the structure of American industry in the postwar period was geared to a previous era (before 1940) in which the United States had a comparative advantage in heavy industrial products that were intensive in complementary inputs of capital and natural resources (including energy). Thus, American firms have been struggling to emerge from being locked into a technological trajectory that left them uncompetitive in a world where comparative advantages in such products have shifted to other nations, as more abundant natural resources were discovered and exploited abroad. In the new era, countries like Japan and West Germany succeeded precisely by developing technologies (both processes and products) that economized on energy and other natural resources. In addition, Nelson and Wright (1992) point out that many U.S. manufacturers had emphasized large-scale mass-production techniques, while high-technology industries benefited from the U.S. lead in education, research, and development. But in the last few decades, trade liberalization and economic integration generally have reduced or eliminated the U.S. special advantage in large-scale production methods, while greater educational and R&D efforts in some other countries (notably Germany and Japan) have enabled them to close the gaps in many high-tech sectors.

All of these factors indicate a reduced ability of the United States to gain monopoly rents from technological innovation at the front end of the product cycle. At the other end, casual empiricism suggests a tremendous speedup in the rate of diffusion or transfer of new technology to low-wage developing countries. The lengthy maturity phase postulated by Vernon (1966) seems to have been squeezed out, as new products quickly become standardized, and their production shifts rapidly to offshore, low-wage areas. Thus, even though the United States retains its technological edge in many product lines, it is increasingly unable to maintain domestic production of the new products (or, at least, of major components of the new products) for very long. Productivity in many NICs' export industries is well over half of U.S. productivity (see, e.g., Blomstrom and Wolff 1994, on Mexico), in spite of wage rates ranging from 10 to 25 percent of U.S. wages (U.S. Department of Labor 1991). The presumption from Ricardian trade theory that international wage differentials merely reflect average productivity differentials—often asserted by economists on a priori grounds—is no longer valid, at least for tradable goods in sectors where capital and technology are mobile (low wages in LDCs may still reflect low productivity in domestic nontradables sectors due to "technological dualism").

In this kind of global competitive environment, U.S. producers have to run faster just to stay in place, and many are giving up the chase (or rather, choosing to run it elsewhere). This character of changing global technological competitiveness can account for why the United States' greatest problem lies in an accelerating structural tendency for imports to increase. In spite of the catch-up of other nations in technological leadership, the United States still retains enough of a lead in enough areas (e.g., aviation equipment, telecommunications, and advanced computers) to keep its exports growing relatively well, provided that foreign income growth is sufficiently rapid and the dollar is not too high. There is only a comparatively slow erosion of the U.S. share of industrial country exports. But the United States is having increasing difficulty in retaining domestic production of existing import-competing products (without sectoral protection, that is) and is even importing more and more relatively high-tech products that it formerly might have exported.

Structural Problems with Particular Countries

The preceding discussion suggests that U.S. structural trade problems are concentrated with two groups of countries. The first group consists of those countries that have caught up to the United States (or are surpassing it) in technological innovation. The second group consists of the countries that have most successfully combined relatively low wages with relatively high productivity in manufacturing assembly operations (usually through some combination of technology transfers and foreign direct investment) to gain overall competitive advantages in unit labor costs in standardized products. Even then, as discussed above, there are macroeconomic adjustment mechanisms that could come into play, but do not do so automatically. Structural trade deficits will emerge, therefore, with those countries in which the requisite macroeconomic adjustments are effectively blocked by state policy or institutional barriers.

A look at the data on bilateral U.S. trade balances in recent years confirms that the most persistent deficits have been with countries that match these descriptions, especially Japan and some of the rapidly industrializing nations of East Asia. Table 2.4 gives data on U.S. merchandise trade balances with the country's leading trading partners and major regions of the world.

TABLE 2.4. U.S. Merchandise Trade Balances by Country, 1980 and 1987–92 (in billions of dollars)

Country	1980	1987	1988	1989	1990	1991	1992
W. Europe	21.1	−27.5	−16.2	−4.0	2.2	14.9	3.2
EC	16.5	−22.0	−11.6	−1.0	4.9	15.6	6.7
Germany	−0.4	−15.4	−12.1	−8.3	−9.7	−5.3	−8.4
Canada	−1.5	−11.6	−10.3	−8.9	−9.6	−7.1	−9.7
Japan	−10.5	−56.9	−52.6	−49.7	−41.8	−44.3	−50.5
Mexico	2.6	−5.7	−2.7	−2.5	−2.4	1.6	4.9
Latin America[a]	1.2	−12.3	−7.8	−8.7	−10.1	0.3	6.2
OPEC	−41.4	−13.7	−9.2	−17.4	−24.8	−15.0	−13.0
China	2.8	−2.8	−3.4	−6.2	−10.4	−12.7	−18.3
Korea	0.2	−9.3	−9.5	−6.7	−4.5	−2.1	−2.8
Taiwan	−2.8	−17.5	−12.8	−14.5	−11.5	−10.3	−10.1
All countries	−25.5	−159.6	−127.0	−115.2	−109.0	−73.8	−96.1

Source: U.S. Department of Commerce, Bureau of Economic Analysis, "U.S. International Transactions," *Survey of Current Business,* June 1993, 78.

[a]Includes all Western Hemisphere countries except Canada.

The data are presented annually for the years 1987 through 1992, with data for 1980 included for purposes of historical comparison. Not surprisingly, when the overall trade deficit peaked in 1987, the United States had deficits with every country and region shown in this table. At that time, one could have attributed most of the deficit to macroeconomic policies and exchange rates, as discussed earlier. By the early 1990s, the overall deficit had been reduced substantially. There were essentially three reasons for this adjustment: the growth slowdown and recession in the United States, the falling value of the dollar, and more rapid growth abroad (at least up to 1991). The countries and regions with which the United States still had deficits in the early 1990s, after all of these macroeconomic adjustments had taken place, are therefore the ones with which we may infer the greatest structural trade problems.

Far and away the largest and most persistent deficit is with Japan, the nation that has clearly offered the greatest challenge to American technological leadership in the postwar period. The dollar fell from 238 yen in 1985 to 128 yen in 1988; the dollar then rose only to 145 yen in 1990 before falling again to 127 yen in 1992.[17] In spite of this massive appreciation of the yen, as well as the fact that the U.S. budget deficit shrank as a percentage of the national product during most of the period from 1985 to 1990, the trade deficit with Japan fell only slightly from $56.9 billion in 1987 to a low of $41.8 billion in 1990, before rising again to $50.5 billion in 1992.

In contrast, the U.S. trade balance with Western Europe went from a deficit of $27.5 billion in 1987 (actually, the deficit with Europe peaked at

$28.6 billion in 1986) to a *surplus* of $14.9 billion in 1991 (which then fell to a $3.2 billion surplus in 1992, as European growth slowed). To be sure, these small surpluses with Western Europe in the early 1990s are less than the surpluses that the United States enjoyed with that region as recently as 1980 and are significantly smaller in proportion to the level of GNP. Nevertheless, it is clear that among the industrialized countries, the Western European nations have borne the lion's share of the adjustment to the global trade imbalances of the United States, while Japan (which had a larger surplus to begin with) has borne relatively little of the adjustment burden in spite of massive currency appreciation.

There is still some room for macro policy correctives with Japan. Surely a massive fiscal stimulus in Japan would not only help to end that country's recession but also would help to reduce that country's record trade surplus (which is global, not just bilateral with the United States).[18] Some have suggested that further yen appreciation might be more effective than the previous appreciation of 1985 through 1988, because Japanese profit margins are now lower and could not be squeezed any further in order to prevent exchange rate pass-through.[19] But it is difficult to believe that these measures will be sufficient, in the absence of correctives to the underlying sources of Japan's structural trade surplus. These sources include Japan's notoriously closed markets for manufactured goods, as well as the aggressive export-oriented strategy of its corporations (including their subsidiaries in other countries) (see Dornbusch 1990; Salvatore 1990; Lawrence 1987, 1991).

Since formal trade liberalization would be largely beside the point, the solution must lie either in some kind of international antitrust agreements with teeth or negotiated "results-oriented" managed trade agreements with allotted foreign shares of Japanese markets and restraints on their exports. The idea of an across-the-board surcharge on manufactured imports from Japan has also been suggested (see Dornbusch 1990; Salvatore 1990), if Japan fails to meet specified market-opening targets.

The other group of countries with chronic large trade surpluses with the United States in recent years has been a somewhat shifting group of low-wage East Asian exporters of manufactures. In 1987, the two largest offenders were South Korea and Taiwan, with which the United States then had a combined deficit of $26.8 billion—nearly as great as the deficit with all of Western Europe at that time! Since then, Korea has made great strides toward rectifying its trade surplus with the United States, helped by significant real depreciation of the won and major increases in domestic wages and aggregate demand. The advent of effective political democracy in Korea probably played an important part in making these adjustments possible, and it shows that they are consistent with domestic economic gains.

In contrast, the deficit with Taiwan (Republic of China) has remained stuck at over $10 billion, and the deficit with the People's Republic of China has come out of nowhere to reach $18.3 billion in 1992. The two Chinas (People's Republic and Taiwan) together had a combined surplus of $28.4 billion in 1992, even larger than the combined Korea-Taiwan surplus of 1987 (when the overall U.S. deficit was more than 50 percent larger), and more than half of the 1992 Japanese surplus with the United States. If we add on the remaining deficits with Korea and Singapore (not shown in the table), the total with the East Asian industrializing countries is well over $30 billion.

The explanation of the huge deficits with China and Taiwan is simple: low unit labor costs due to the combination of low wages (actually, much lower in China than in Taiwan) with high productivity in modern industries (but productivity is also higher in Taiwan, undoubtedly). To a trade theorist, this is a disequilibrium situation with a simple and natural remedy: if wages in China and Taiwan do not rise sufficiently, revaluation of the Chinese and Taiwanese currencies can raise their average unit labor costs in dollars and eliminate their overall competitive advantage. But the problem is that there is no automatic mechanism in the international financial marketplace to compel this outcome. Both countries have heavily managed exchange rate regimes. Both nations' central banks have amassed enormous reserves as they have bought up excess supplies of foreign exchange in order to keep their currencies undervalued.[20] And both nations have authoritarian political regimes that effectively keep the lid on domestic wages.

In this situation, unless the democratization movements in these countries succeed, the only solution is outside political pressure. The need to renew China's most favored nation trading status with the United States could be used as an occasion to obtain commitments in this regard, including revaluing the yuan against the dollar. It is important to include a specific economic demand like this in our negotiations with the Chinese government, as well as the broader human rights and political democracy concerns. While the latter are extremely important, as a practical matter we are much more likely to get the Chinese government to adjust its exchange rate than to change its political system. Similar pressures should be applied to Taiwan. In both China and Taiwan, political liberalization would undoubtedly help with trade adjustment, as industrial workers would claim higher wages consistent with their rising productivity, and citizens could win more consumption-oriented economic policies, both of which would help to reduce trade surpluses. But revaluing undervalued currencies is something that can be demanded even of current political regimes.

The question then arises of what to do if these countries do not raise their average unit labor costs in dollars, either through wage increases or currency appreciation, to levels consistent with balanced trade. In Blecker (1992a), I have urged the threatened imposition of a unit labor cost equalization surcharge on manufactured imports from countries like China and Taiwan, which have large and persistent surpluses accounted for by low unit labor costs in dollar terms and a record of exchange-rate intervention to keep their currencies undervalued.[21] Clearly, revaluing the currency should be a preferable option for these countries, compared with accepting a surcharge, since the revaluation brings improved terms of trade while the surcharge does not. Thus, with revaluation there is a gain in purchasing power over imports that offsets the lessened competitiveness of exports; with the surcharge there is no such offsetting gain. It is to be hoped, then, that countries like China and Taiwan would become willing to revalue in order to avoid actual imposition of the surcharge.

The issue of market openness also arises in regard to China and Taiwan. Both countries have maintained tight effective controls over the importation of manufactured goods that could compete with nascent domestic industries. Taiwan has used its import controls to create a hothouse environment favorable to the growth of small domestic firms, while China is now striking deals with foreign multinationals like McDonnell-Douglas to acquire state-of-the-art technology in advanced industries. While these countries have relied on open markets abroad, particularly in the United States, to fuel their export-led growth, they have not been willing to reciprocate by opening their home markets. This is mercantilism, pure and simple, and it has no place in a truly "liberal" trading system.

The United States, with its declining real wages and sluggish growth, can no longer afford to keep its market unilaterally open in exchange for geopolitical or strategic (but noneconomic) advantages. This is especially true in the case of the two Chinas, whose rivalry is a relic of the Cold War era and whose ideological disputes are increasingly just a cover for two authoritarian regimes that have more in common than they would like to admit. What U.S. tolerance of their undervalued exchange rates and closed markets does is only to prop up these undemocratic states at the expense of American workers who lose jobs to artificially cheap imports. Even with balanced trade, of course, we would continue to import large amounts of labor-intensive manufactures from China, Taiwan, and other East Asian NICs. But the current levels of imports go far beyond what can be justified by the principle of comparative advantages due to the enormous trade imbalances. In 1992, China's exports to the United States were *more than three times* the level of its imports from the U.S. ($25.7 vs. $7.5 billion); Taiwan's exports to the United States were *nearly double* its imports ($24.6 vs. $14.5 billion). This is not Ricardian comparative advantage trade!

Conclusion

The discussion of the Pacific Rim surplus countries in the previous section points out the subtleties involved in trying to disentangle structural and macroeconomic causes of trade imbalances. Since Japan, China, and Taiwan together account for roughly 80 percent of the 1992 U.S. trade deficit, these three cases are clearly consequential to the aggregate outcome. In each case, it is possible to identify structural factors that contribute to large trade surpluses, including closed internal markets and highly competitive exports (Japan more in terms of rising technological sophistication and high product-quality, China and Taiwan more in terms of low unit labor costs). Nevertheless, there are also macroeconomic factors at work, including undervalued currencies and insufficiently expansionary fiscal and monetary policies. In the cases of these countries, the structural and macroeconomic determinants of trade balances seem to be reinforcing each other in creating chronic surpluses, rather than offsetting each other and eliminating surplus tendencies. Macroeconomic correctives could be made, but are impeded by political institutions and entrenched interests. Market forces alone do not automatically call forth such correctives.

But as noted earlier, even allowing macroeconomic correctives to work can have its costs. Depreciating the dollar lowers the U.S. terms of trade and the purchasing power of domestic incomes. Slower income growth in the United States helps to keep the trade balance down but at tremendous social cost. For these reasons, it is imperative for the surplus countries to shoulder more of the burden of adjustment. Expansionary macro policies and market opening can both raise domestic living standards and cure external imbalances in surplus countries that are suppressing consumer welfare. This is the growth path, rather than the stagnation path, to restoring global trade balances.

In the end, our discussion shows that the structural and macroeconomic aspects of trade balances are closely intertwined. The structural characteristics (e.g., labor productivity, trade barriers, product quality, technological innovation) determine the parameters within which traditional macro policies (fiscal, monetary, and exchange rate) operate, while the latter determine the degree to which the structural factors are manifest in trade imbalances or in other variables. To ignore the macroeconomic relationships and policies would be naive and misleading. But it is equally wrong to ignore the structural causes of trade imbalances, or to pretend that they are automatically and costlessly offset by macroeconomic adjustments. The precise way in which structural and macroeconomic factors interact will vary over time and between countries. Understanding the variety of ways in which they can interact and the full range of policy alternatives for dealing with imbalances when they arise is essential for sound policy formulation.

Appendix

The new estimates of income elasticities for U.S. exports and imports are based on a model similar to that used by the author in Blecker (1992b). The main difference is that the new estimates make an adjustment for the overvaluation of the "real" volume of computers in the official U.S. data. Those data, as published by the U.S. Department of Commerce, Bureau of Economic Analysis in the National Income and Product Accounts (NIPAs), measure the "hedonic" (qualitative) attributes of computers at constant, base-year prices. Given the rapid technical progress in computers in the past decade, this procedure results in an exaggerated picture of the volume of computer imports and exports. Essentially, the NIPA approach emphasizes the amount of computing power traded, rather than the number of computers. This is inconsistent with how all other goods in the NIPA are treated.

TABLE 2A.1. Econometric Estimates of Demand Functions for U.S. Nonagricultural Exports and Nonpetroleum Imports with Alternative Measure of Computers (sample period: 1978-I to 1990-IV)

Variable	Imports	Exports
Constant	12.62	3.28
	(28.07)	(5.03)
Relative price[a]	1.27	0.73
	(9.48)	(15.68)
Income[b]	2.22	1.38
	(40.43)	(22.15)
Adjusted R^2	0.99	0.96

Source: Author's calculations.

Note: The dependent variables are (respectively) nonagricultural merchandise exports and nonpetroleum merchandise imports in constant 1982 dollars, except for exports and imports of computers, which were revalued at current prices. All variables are measured in natural logarithms. All equations were estimated by ordinary least squares using quarterly data. Numbers in parentheses are *t*-statistics.

[a]For exports, this is the implicit price deflator for nonagricultural, noncomputer exports multiplied by the Federal Reserve Board index for the real value of the dollar and divided by the U.S. consumer price index (so as to obtain an index of U.S. export prices relative to a trade-weighted average of foreign price levels). For imports, this is the implicit price deflator for nonpetroleum, noncomputer imports relative to producer price index for nonfuel industrial commodities. Coefficients shown in the table are the sums for quarterly lags 0-8.

[b]For exports, the income variable is total GDP for all non-U.S. OECD countries, in constant 1985 prices, converted at 1985 purchasing power parity exchange rates. For imports, the income variable is U.S. gross domestic purchases in billions of 1982 dollars.

The alternative method chosen here was to value computers at their nominal prices in the "real" import and export series, which is equivalent to giving them a price deflator of 1.0. This is the actual practice in the German national accounts. This is essentially an agnostic procedure that admits that we do not have a very good way of measuring the real quantity of computers in a period when their quality is rapidly improving. In effect, the method used here treats a computer as a computer, regardless of when it was produced. The econometric results thus obtained are presented in table 2A.1.

In addition to the regressions reported in table 2A.1, a number of alternative specifications were estimated with the same data set in order to test the sensitivity of the results to the assumptions about computer prices. Using the official data results in higher estimates of both import and export income elasticities, very close to the estimates from Blecker (1992b) reported in table 2.3 in the text above. On the other hand, omitting computers entirely results in lower estimates of both. But in all cases the income elasticity is notably higher for imports.

NOTES

1. All trade balance data in this paragraph are on a balance-of-payments basis, as reported in U.S. Congress, Joint Economic Committee (May 1994).

2. See, e.g., Hatsopoulos, Krugman, and Summers 1988 for a concise statement of this view.

3. Strictly speaking, the following equation applies to the *current account* balance, which equals the merchandise trade balance plus the services balance (including net inflows of investment income) and net transfers. In this context, the term *trade balance* should be understood to mean the current account.

4. The definition of macroeconomic stability used here refers to the output adjustment process in the goods market. Assuming that output increases when there is excess demand for goods and decreases when there is excess supply, then output will converge to its equilibrium level if and only if the propensity to save is greater than the propensity to invest out of additional income.

5. In particular, it is important to have a competitive domestic capital goods sector so that the import coefficient of investment is relatively low.

6. For an exposition of this analysis see Bosworth 1993.

7. The Helkie-Hooper estimate of the effect of U.S. fiscal expansion is actually based on a rather crude extrapolation of the simulation results for the MCM model reported in the Bryant et al. comparison of models. Those results show a sustained 1 percent of GNP fiscal expansion in the United States causing the U.S. current account to fall by about $10 billion after one year, $17 billion after two years, $23 billion after three years, and $53 billion after six years. Helkie and Hooper state that, based on these results, they "have chosen to estimate" the current account effect of U.S. fiscal expansion

as a fall of $20 billion for each 1 percent of GNP of fiscal stimulus—apparently, the effect after two and a half years (since 20 is halfway between 17 and 23). Then they multiplied this figure by 3.5 (the approximate percentage of GNP by which the U.S. fiscal deficit rose in 1980–85) to obtain the estimate of a $70 billion fall in the current account. It should be noted that this procedure exaggerates the effects of the actual shift in U.S. fiscal policy, since some part of the 3.5 percent of GNP increase in the actual U.S. federal deficit must be attributed to the foreign demand contraction combined with U.S. competitive decline.

8. In addition, they estimated that "foreign fiscal contraction contributed another $25 billion to the deficit," so that domestic and foreign fiscal policies combined explain a total of $95 billion of the increased current deficit, or about two-thirds of the total increase (Helkie and Hooper 1988, 49). Most of the remainder of this increase is attributed to the "additional" dollar appreciation in 1980 through 1985—i.e., the part of the dollar's rise that is not explained by fiscal policies (and is presumably due to monetary contraction plus the speculative bubble).

9. Note that the trade deficit peaked in 1987, two years after the peak in the real value of the dollar, due to the lagged effects of exchange rate changes on volumes of trade—the well-known J-curve effect.

10. The other nations shown are most of the United States' leading competitors, including all of the other G-7 countries plus Korea representing the newly industrializing countries (NICs). Other major NICs (such as Taiwan, Brazil, and Mexico) were omitted due to data problems.

11. It could be argued that the nation nevertheless kept its consumer standard of living rising by living beyond its means through public sector dissaving and private sector overconsumption (see, e.g., Hatsopoulos, Krugman, and Poterba 1989). While there is some tautological truth to this in regard to the federal budget deficit, the notion of a consumption binge by most private households has clearly been exaggerated (see Blecker 1990). It is also true that households have been able to keep their living standards from falling by having household members work more hours in order to compensate for falling hourly real wages. But the need to work more hours is itself evidence of a decline in the overall standard of living, including leisure time.

12. These series are taken from the national income and product accounts (NIPAs) and thus differ slightly from the international transactions (balance of payments) series. The current account is measured by the NIPA concept of net foreign investment.

13. The Fed index is used since it goes back two years earlier than the IMF index, which starts in 1975. The trend is calculated by a least-squares regression of the natural logarithm of the exchange rate index on the time period for the years 1973–79, with the rest of the trend projected out-of-sample.

14. For further discussion of this issue see Meade 1991 and Blecker 1992a.

15. See the appendix for details on these new estimates.

16. Of course, if incomes do not adjust, then persistent trade imbalances will emerge, sustained by international capital movements (lending from surplus countries to deficit countries). This is what has happened to the United States in the last decade.

17. Averages of daily exchange rates from the U.S. Board of Governors of the Federal Reserve System, as published in U.S. Council of Economic Advisors (1993, table B-107, 470).

18. According to the International Monetary Fund (*International Financial Statistics*, March 1993, (306–7), Japan had an overall merchandise trade surplus of $103

billion in 1991 and was running quarterly trade surpluses of over $30 billion for the first three quarters of 1992 (indicating an annual surplus approaching $125 billion for 1992).

19. Conversation with Rudiger Dornbusch, Anaheim, California, January 1993.

20. As of mid-1992, China's reserves totaled over $31 billion, while Taiwan's reserves surpassed $61 billion. These are among the largest reserves of any country in the world and are much larger than those of most other developing countries (only Singapore comes close, with reserves of $26 billion in mid-1992). See International Monetary Fund, *International Financial Statistics,* March 1993, 29–31.

21. Note that this surcharge is designed only to be applied in these types of cases, not in all instances of low wages. It is not, therefore, what is sometimes called a scientific tariff to equalize *wages* per se, and it would not be applied in any case to countries without undervalued currencies *and* chronic trade surpluses vis-à-vis the United States.

REFERENCES

Bernheim, B. Douglas. 1988. Budget Deficits and the Balance of Trade. *Tax Policy and the Economy* 2:1–31.

Blecker, Robert A. 1990. *Are Americans on a Consumption Binge? The Evidence Reconsidered.* Washington, D.C.: Economic Policy Institute.

———. 1991a. Low Saving Rates and the Twin Deficits: Confusing the Symptoms and Causes of Economic Decline. In *Economic Problems of the 1990s,* ed. Paul Davidson and Jan A. Kregel. Aldershot (U.K.) and Brookfield, Vt.: Edward Elgar.

———. 1991b. *The Recession, the Dollar, and the Trade Deficit: Recent Trends and Future Prospects.* Economic Policy Institute Briefing Paper. Washington, D.C.: Economic Policy Institute.

———. 1992a. *Beyond the Twin Deficits: A Trade Strategy for the 1990s.* Armonk, N.Y.: M. E. Sharpe, Inc., Economic Policy Institute Series.

———. 1992b. Structural Roots of U.S. Trade Problems: Income Elasticities, Time Trends, and Hysteresis. *Journal of Post Keynesian Economics* 14, no. 3: 321–46.

———. ed. 1996. *U.S. Trade Policy and Global Growth: New Directions in the International Economy.* Armonk, N.Y.: M. E. Sharpe, Inc., Economic Policy Institute Series.

Block, Fred C. 1990. Bad Data Drive Out Good: The Decline of Personal Savings Reconsidered. *Journal of Post Keynesian Economics* 13, no. 1: 3–19.

Blomstrom, Magnus, and Edward N. Wolff. 1994. Multinational Corporations and Productivity Convergence in Mexico. In *Convergence of Productivity: Cross-National Studies and Historical Evidence,* ed. William J. Baumol. Oxford: Oxford University Press.

Bosworth, Barry P. 1993. *Saving and Investment in a Global Economy.* Washington, D.C.: The Brookings Institution.

Bryant, Ralph C., et al. 1988. *Empirical Macroeconomics for Interdependent Economies.* Washington, D.C.: The Brookings Institution.

Cline, William R. 1989. *United States External Adjustment and the World Economy.* Washington, D.C.: Institute for International Economics.

Cooper, Richard. 1987. Symposium on the Causes of the U.S. Trade Deficit. United States General Accounting Office, Report to Congressional Requesters.

Dixit, Avinash. 1992. Investment and Hysteresis. *Journal of Economic Perspectives* 6, no. 1: 107–32.

Dornbusch, Rudiger. 1988. The Adjustment Mechanism: Theory and Problems. In *International Payments Imbalances in the 1980s,* ed. Norman S. Fieleke. Boston: Federal Reserve Bank of Boston.

———. 1989. The Dollar, U.S. Adjustment and the System. Paper presented at American Economic Association Meetings, Atlanta, December.

———. 1990. Is There a Case for Aggressive Bilateralism and How Best to Practice It? In *An American Trade Strategy: Three Options for the 1990s,* ed. Robert Z. Lawrence and Charles L. Schultze. Washington, D.C.: The Brookings Institution.

Dornbusch, Rudiger, Paul Krugman, and Yung Chul Park. 1989. *Meeting World Challenges: U.S. Manufacturing in the 1990s.* Rochester, N.Y.: Eastman Kodak Company.

Dosi, Giovanni, Keith Pavitt, and Luc Soete. 1990. *The Economics of Technical Change and International Trade.* New York: New York University Press.

Eisner, Robert. 1991. The Real Rate of U.S. National Saving. *Review of Income and Wealth,* series 37 (March).

Fazzari, Steven M. 1993. *Investment and U.S. Fiscal Policy in the 1990s.* Briefing Paper, June. Washington, D.C.: Economic Policy Institute.

Fazzari, Steven M., and Michael J. Athey. 1987. Asymmetric Information, Financing Constraints, and Investment. *Review of Economics and Statistics* 69:481–87.

Fazzari, Steven M., R. Glenn Hubbard, and Bruce C. Petersen. 1988. Financing Constraints and Corporate Investment. *Brookings Papers on Economic Activity* 1:141–206.

Frankel, Jeffrey A. 1990. The Making of Exchange Rate Policy in the 1980s. Working Paper no. 3539, National Bureau of Economic Research, December.

Hatsopoulos, George, Paul Krugman, and James Poterba. 1989. *Overconsumption: The Challenge to U.S. Policy.* New York and Washington, D.C.: American Business Conference.

Hatsopoulos, George, Paul Krugman, and Lawrence Summers. 1988. U.S. Competitiveness: Beyond the Trade Deficit. *Science* 241:299–307.

Helkie, William L., and Peter Hooper. 1988. An Empirical Analysis of the External Deficit. In *External Deficits and the Dollar: The Pit and the Pendulum,* ed. Ralph C. Bryant et al. Washington, D.C.: The Brookings Institution.

Hooper, Peter. 1989. U.S. Net Foreign Saving Has Also Plunged. *Challenge* (July–August): 33–38.

Hooper, Peter, and Catherine L. Mann. 1989. The U.S. External Deficit: Its Causes and Persistence. In *U.S. Trade Deficit: Causes, Consequences, and Cures,* ed. Albert E. Burger. Proceedings of the Twelfth Annual Economic Policy Conference of the Federal Reserve Bank of St. Louis (1987). Boston: Kluwer Academic Publishers.

International Monetary Fund (IMF). Various years. International Financial Statistics. Washington, D.C.: IMF.

Krugman, Paul R. 1979. A Model of Innovation, Technology Transfer, and the World Distribution of Income. *Journal of Political Economy* 87, no. 2: 253–66.

———. 1989a. *Exchange-Rate Instability.* Cambridge, Mass: MIT Press.

———. 1989b. Differences in Income Elasticities and Trends in Real Exchange Rates. *European Economic Review* 33:1031–54.

———. 1991. *Has the Adjustment Process Worked?* Policy Analyses in International Economics, no. 34. Washington, D.C.: Institute for International Economics.

Krugman, Paul R., and Richard E. Baldwin. 1987. The Persistence of the U.S. Trade Deficit. *Brookings Papers on Economic Activity,* no. 1: 1–43.

Lawrence, Robert Z. 1987. Imports in Japan: Closed Markets or Minds? *Brookings Papers on Economic Activity,* no. 2: 517–54.

———. 1989. The International Dimension. In *American Living Standards: Threats and Challenges,* ed. Robert E. Litan et al. Washington, D.C.: The Brookings Institution.

———. 1990. U.S. Current Account Adjustment: An Appraisal. *Brookings Papers on Economic Activity,* no. 2: 343–89.

———. 1991. Efficient or Exclusionist? The Import Behavior of Japanese Corporate Groups. *Brookings Papers on Economic Activity,* no. 1: 311–41.

Levich, Richard M. 1988. Financial Innovations in International Financial Markets. In *The United States in the World Economy,* ed. M. Feldstein. Chicago: University of Chicago Press/NBER.

McCauley, Robert N., and Steven A. Zimmer. 1989. Explaining International Differences in the Cost of Capital. *Federal Reserve Bank of New York, Quarterly Review* (summer): 7–28.

McCulloch, Rachel. 1986. Trade Deficits, Industrial Competitiveness, and the Japanese. *California Management Review* 27, no. 2: 140–56; reprinted in *International Trade and Finance: Readings,* ed. Robert E. Baldwin and J. David Richardson, 3d ed. Boston: Little, Brown.

McCulloch, Rachel, and J. David Richardson. 1986. U.S. Trade and the Dollar: Evaluating Current Policy Options. In *Current U.S. Trade Policy: Analysis, Agenda, and Administration,* ed. Robert E. Baldwin and J. David Richardson. Cambridge, Mass.: NBER.

McKibbin, Warwick J., and Jeffrey D. Sachs. 1989. The McKibbin-Sachs Global Model: Theory and Specification. National Bureau of Economic Research. Working Paper no. 3100.

Meade, Ellen E. 1991. Computers and the Trade Deficit: The Case of the Falling Prices. In *International Economic Transactions: Issues in Measurement and Empirical Research,* ed. Peter Hooper and J. David Richardson. Chicago: University of Chicago Press.

Nelson, Richard R., and Gavin Wright. 1992. The Rise and Fall of American Technological Leadership: The Postwar Era in Historical Perspective. *Journal of Economic Literature* 30, no. 4: 1931–64.

Organisation for Economic Co-operation and Development (OECD). 1991. *Main Economic Indicators.* Paris: OECD.

Rosensweig, Jeffrey A., and Ellis W. Tallman. 1993. Fiscal Policy and Trade Adjustment: Are the Deficits Really Twins? *Economic Inquiry* 31: 580–94.

Sachs, Jeffrey D. 1988. Global Adjustments to a Shrinking U.S. Trade Deficit. *Brookings Papers on Economic Activity* 2: 639–74.

Salvatore, Dominick. 1990. *The Japanese Trade Challenge and the U.S. Response: Addressing the Structural Causes of the Bilateral Trade Imbalance.* Washington, D.C.: Economic Policy Institute.

Thirlwall, A. P. 1979. The Balance of Payments Constraint as an Explanation of International Growth Rate Differences. *Banca Nazionale del Lavoro Quarterly Review* 128: 45–53.

U.S. Congress, Joint Economic Committee. Various issues. *Economic Indicators.*

U.S. Council of Economic Advisors. Various years. *Economic Report of the President.* Washington, DC: Government Printing Office.

U.S. Department of Commerce, Bureau of Economic Analysis. 1986. *National Income and Product Accounts, 1929–82.* Washington, D.C.: Government Printing Office.

———. Various issues. *Survey of Current Business.*

U.S. Department of Labor, Bureau of Labor Statistics (BLS). 1991. "International Comparisons of Hourly Compensation Costs for Production Workers in Manufacturing, 1990." Report 803, May.

U.S. National Science Foundation. 1990. *National Patterns of R&D Resources, 1990.* Report 90–316.

Vernon, Raymond. 1966. International Investment and International Trade in the Product Cycle. *Quarterly Journal of Economics* 80:190–207.

Wright, Gavin. 1990. The Origins of American Industrial Success, 1879–1940. *American Economic Review* 80, no. 4: 651–68.

CHAPTER 3

Improving U.S. International Competitiveness

Macro Policy Management vs. Managed Trade Policy

Catherine L. Mann

A classic debate among economists and policymakers is under way on how to improve U.S. competitiveness, particularly as measured in the international arena. Macroeconomists argue that U.S. international competitiveness is a function of savings and investment and the mix of fiscal and monetary policy. Thus, macroeconomic policy management is the key to improving competitiveness. Microeconomists suggest that incentives and structures in the business environment facing U.S. industry are the fundamental determinants of U.S. competitiveness. Therefore, the policy effort should focus on improving the environment for business. To what extent are these two views complementary, and to what extent do they lead to similar or different policy prescriptions? What are the channels of transmission of either macroeconomic or microeconomic policies to international competitiveness?

The behavior of firms is a key link between macro and micro policy prescriptions to improve international competitiveness. For example, the macroeconomic policy mix will affect business prices and profits through its effect on the level of economic activity, interest rates, inflation, and the exchange rate. On the other hand, trade policies and domestic industrial policies may affect market structure and market access, which will influence prices and profits of firms and may affect their opportunity to compete in international markets.

Looking at the experience of the United States over the last decade, what major factors have influenced international competitiveness? Are there lessons on what policies might improve U.S. international competitiveness into the twenty-first century?

The next section addresses why international competitiveness matters and considers alternative measures of it. Then I present a stylized model of the firm and discuss channels through which macro and micro policies might affect firm performance and international competitiveness. The following section looks at determinants of international competitiveness of the United States over the last two decades. I conclude by reviewing the evidence on which macro or micro policies are likely to have the greatest effect on international competitiveness in the next decade.

International Competitiveness: Does It Matter and Can We Measure It?

Why Does International Competitiveness Matter?

International competitiveness matters because it has an important impact on the standard of living. As known from simple growth theory, increases in the standard of living come from increases in the amount of or productivity of resources. Improving international competitiveness raises the standard of living through both channels.

The upper panel of figure 3.1 is the familiar gains-from-trade diagram. Compared to autarky, balanced international trade at terms of trade P_X/P_M increases utility from U_0 to U_1 by concentrating production in the relatively more efficiently produced product X and trading that for the relatively more expensive (in autarky) and inefficiently produced product M. From the initial trading equilibrium U_1, if the terms of trade improves, a permanently higher level of utility U_2 can be achieved using the same amount of home resources. Thus, international trade and improved international competitiveness with balanced trade raise the standard of living.

Second, improved international competitiveness is consistent with an increase in the productivity of domestic resources. The lower panel shows the original and a new production possibility frontier drawn assuming greater productivity of both capital and labor in the production of both M and X. The higher level of productivity yields a higher level of utility U_4, without any change in relative prices. To attain the higher level of utility U_4 with the original level of productivity and production, the trade balance would have to go into deficit, as shown by the parallel lines tangent to the original and higher-productivity PPF. Utility U_4 attained with a trade deficit cannot be maintained permanently since the deficit will at some point have to be rectified, at a cost to consumption and utility. But, if the underlying productivity of factors increases, as described by the outward movement of the PPF from its original location, the trade deficit supporting utility level U_4 can be closed without a reduction in the standard of living, and the country can maintain this higher utility permanently.

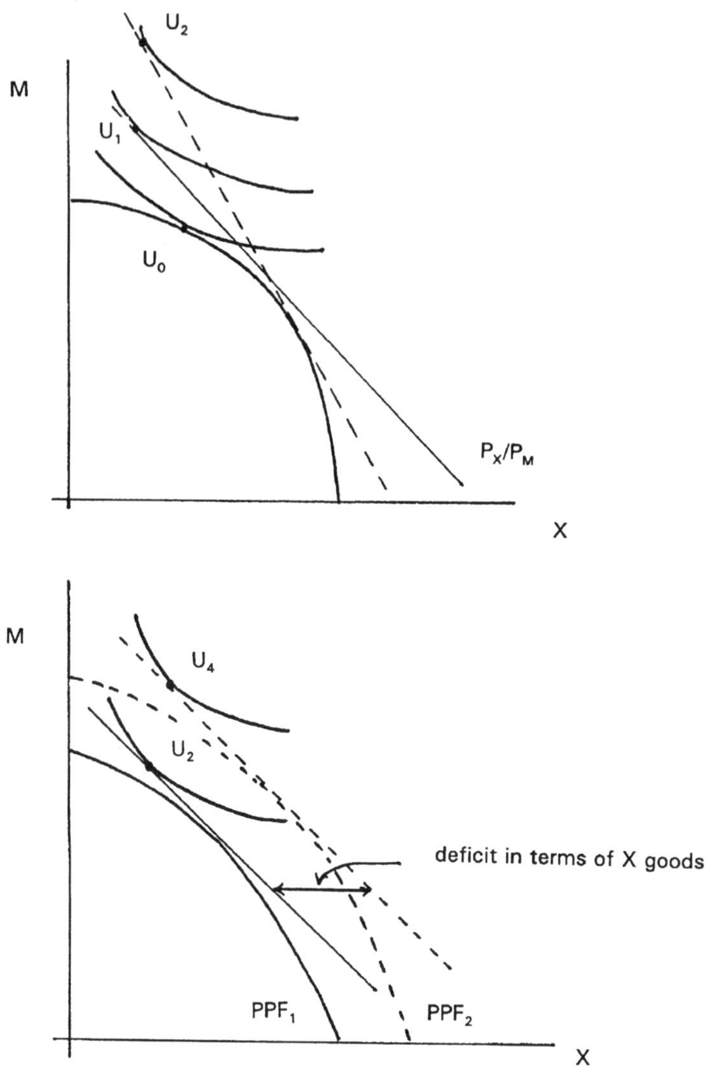

Fig. 3.1. International competitiveness and the standard of living: a theoretical presentation

How To Measure International Competitiveness

Before investigating measures of international competitiveness, it is useful first to determine where we are with respect to our standard of living and how it has been changing over time, especially compared with other nations. A common measure of the standard of living is real GDP per person, shown in the upper panel of figure 3.2 for the United States, Germany, Japan, and Korea

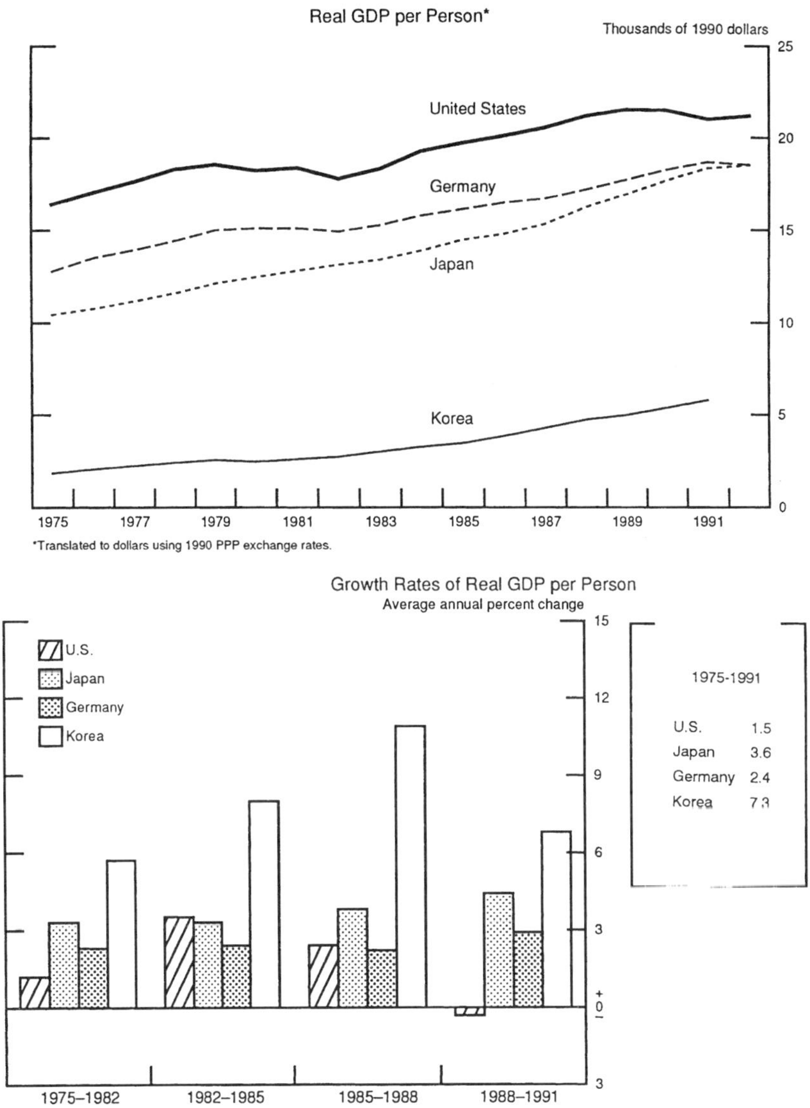

Fig. 3.2. Measures of the standard of living. (Adapted from Bureau of Labor Statistics, unpublished data.)

from 1975 to 1992. Germany and Japan are shown because they are significant trading partners as well as countries whose products compete with U.S. exports in international markets. Korea is chosen as broadly representative of the fast-growing and increasingly internationally important exporters and markets of East Asia.

The United States continues to have the highest standard of living, as measured by real GDP per person, although that level did decline between 1989 and 1991. Germany and Japan now have similar standards of living, although the Japanese standard of living grew much faster over the last 20 years. For these two countries, as for the United States, the standard of living has stagnated in the 1990s. Korea's estimated real GDP per person has doubled in the last 20 years, although the level is well below that of three industrial countries.

The trends shown in the upper panel appear even more striking in the bar chart in the lower panel. Korea has doubled its per-person real GDP with real growth rates averaging between 6 and 11 percent per year over the nearly 20-year period. Japan's standard of living rose to that of Germany's by increasing only 1 or 2 percent per year faster. The growth rate of per-person GDP in the United States has been much more variable over the period, at times outstripping that of Germany and Japan (1982–85), but more frequently lagging.

The analytical example suggested two measures of international competitiveness: the terms of trade and the trade balance. Figure 3.3 shows these measures of international competitiveness for the United States along with the real exchange rate for the last 20 years. As measured by the trade balance, U.S. international competitiveness worsened between 1975 and 1977, recovered to 1980, then declined precipitously between 1981 and 1986. Between 1987 and 1991, the trade balance recovered, but then resumed declining. As measured by the terms of trade, U.S. international competitiveness worsened between 1975 and 1980, improved modestly through 1985, declined sharply to 1986, and has drifted downward subsequently. The real exchange rate has followed a similar pattern. The point to draw from this simple examination is that these two measures of international competitiveness sometimes move together and sometimes move in opposite directions.

How do these measures of international competitiveness fit into the macro-micro debate? The macroeconomist might be drawn to the trade balance measure since the export-import balance is related to the savings-investment and domestic absorption-expenditure balances that are at the heart of macroeconomic analysis. The microeconomist might be drawn to the terms of trade measure since the determinants of export and import prices, or more generally price theory, are at the core of microeconomics.

In the analytical example, the effect of a change in relative prices and the trade balance were examined independently. Figure 3.4 indicates clearly their interdependence. The figure shows the actual nonoil, nonagricultural trade balance since 1980, and a decomposition of this partial trade balance into two components: the portion of the deficit attributable to changes in the exchange rate and import and export prices (the dotted area) and the portion of the

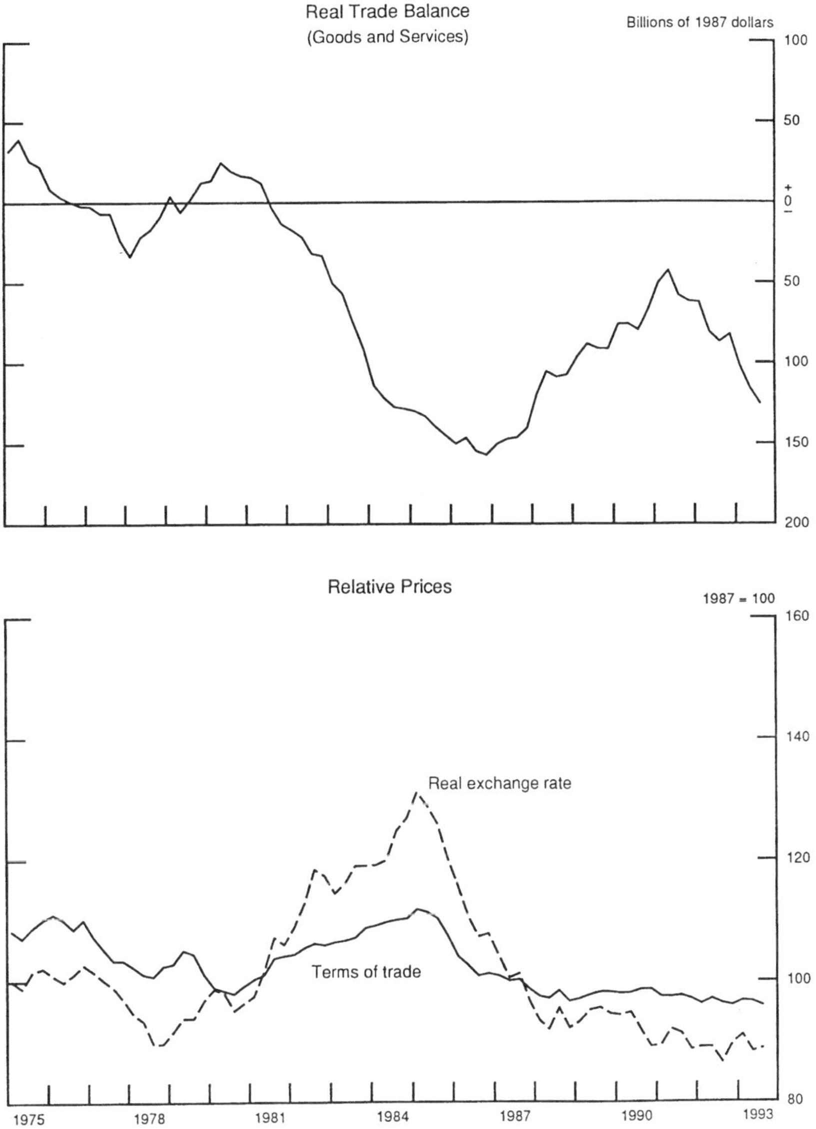

Fig. 3.3. Measuring international competitiveness. (Author's calculations from the Federal Reserve Board.)

trade deficit attributable to changes in income growth differentials between the United States and its trading partners (the hatched area).[1]

The decomposition emphasizes several points. First, the portion of the deficit that remains unexplained after accounting for changes in incomes,

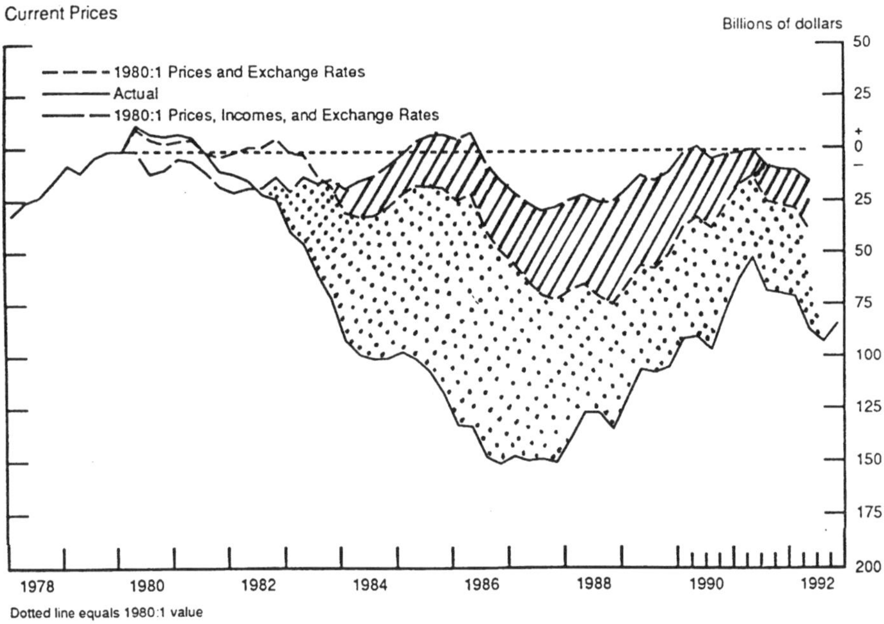

Fig. 3.4. Counterfactual simulations for U.S. partial trade balance. (Adapted from Peter Hooper and Jaime Marquez, "Exchange Rates, Prices, and External Adjustment in the United States and Japan," prepared for the Fiftieth Anniversary of Essays in International Finance, Princeton University, April 1993. Reprinted as the Board of Governors of the Federal Reserve Board, International Finance Discussion Papers no. 456, October 1993.)

relative prices, and exchange rates is relatively small. Thus, understanding the behavior of these three factors essentially explains the evolution of the trade balance since 1980. Second, comparing the relative importance of these factors, changes in the exchange rate and relative prices together were more important than changes in income growth differentials in explaining the evolution of the trade balance between 1980 and 1987. Changes in the exchange rate and relative prices and growth differentials were about equal in importance between 1987 until 1990, when changes in the exchange rate and relative prices reemerged as the dominant proximate determinants of the trade balance. Figure 3.4 makes clear that understanding changes in the international competitiveness of U.S. products, as measured by the exchange rate and relative prices, will go a long way toward understanding changes in international competitiveness as measured by the trade balance.

The Firm: Linking Macro and Micro Policies for Competitiveness

Figure 3.4 emphasized that understanding why the exchange rate and relative prices move is key to understanding changes in international competitiveness as measured by the trade balance. Substantial research time has been devoted to understanding exchange rate behavior without much luck; but, it is fair to say, the fiscal and monetary policy mix as proxied by movements in the real interest rate is a key determinant of the exchange rate.

The point of this paper is that economic policies can also affect relative prices and external balance through a firm's costs, pricing strategy, and profits. A behavioral specification of the NIPA identities and a simple model of price determination shows how the macroeconomic policy mix and microeconomic trade policies can be transmitted through the firm to yield changes in international competitiveness.

A Simple Model of the Firm and Measures of International Competitiveness

A simple and frequently used macroeconomic framework for analyzing external balance starts with the national income and product accounts (NIPA) wherein output (Y) is the sum of consumption (C), investment (I), government spending (G), and net exports (NX). As is well known, an excess of domestic expenditure ($E = C + I + G$) over output (Y) is associated with an excess of imports (M) over exports (X): $Y - E = NX = X - M$. That is, an internal deficit is mirrored by an external deficit.

One simple behavioral model of external balance is a partial equilibrium specification of the determinants of exports and imports. In this partial equilibrium model of trade, exports are a function (x) of foreign income (Y^*) and the relative price of exports (P_x) to competing products (P^*). Imports are a function (m) of domestic income (Y), and the relative price of imports (P_m) to domestic products (P). These observations yield a standard behavioral specification of external balance:

$$X - M = x\,(P_x/P^*, Y^*) - m\,(P_m/P, Y). \tag{1}$$

This formulation shows the relationship between the relative price and the trade balance measures of international competitiveness. Income effects held aside, factors that improve the price competitiveness of export (P_x)

and/or import-competing (P) goods will improve the external balance measure of international competitiveness.

Suppose the firm in this example enjoys some degree of market power so that it can set price to maximize profits. In this general case, price (P) is determined by unit cost (C) and markup (V). Unit cost is a function (c) of a number of factors including labor compensation (w), cost of capital (r), intermediate input costs (m), and production level (Q). The markup is a function (v) of elasticity of demand (e), market structure (n), and time horizon of the firm (t).

$$P = c(w, r, m, Q) * v(e, n, t) \tag{2}$$

To put this general pricing formula in the context of international competitiveness, consider the relative prices in equation (1). For the import-competing firm, the relevant measure of competitiveness is the ratio of the import price (P_m) to the domestic price (P), in a common currency. Thus, the foreign currency price of the imported good (P_f) must be translated into the domestic currency using the exchange rate (xr) quoted in terms of home currency per unit of foreign currency.

$$P_m/P = (P_f^* xr) / P \tag{3}$$

On the export side, the relevant measure of international competitiveness is the ratio of the export price, in foreign-currency units, to the foreign-currency price of competing products (P^*).

$$P_x/P^* = (P_x/xr) / P^* \tag{4}$$

Combining equations (1) through (4) and suppressing the income terms (Y and Y^*)[2] yields the following expression that focuses on the domestic firm and allows for different cost and markup functions (as noted by subscripts) for the exporter and import-competing producer:

$$\begin{aligned} X - M = x\ [(c_X(w, r, m, Q) * v_X(e, n, t))/xr]/P^*) \\ - m\ [P_f^* xr/(c_M(w, r, m, Q) * v_M(e, n, t))]. \end{aligned} \tag{5}$$

Income and foreign price effects held aside, international competitiveness, as measured by either the relative price or by external balance, will be determined by domestic costs, the markup strategy, and the exchange rate.[3]

Channels of Transmission of Policies to International Competitiveness

Equation (5) puts the firm at the center of how macro and micro policies can affect international competitiveness. Fiscal and monetary policy are the tools of macroeconomic policy. The mix of these policies affects the firm's costs and competitiveness through the interest rate, inflation, and exchange rate. The tools of microeconomic policy are harder to characterize so simply, but include market structure policies and trade policies. These policies might affect the ease of market entry and exit, the degree of foreign competition at home, and extent of foreign market access abroad; all of these affect the markup and may also affect costs if economies of scale are important.

With such a simple framework, it is relatively easy to catalog the effects of macro policies on international competitiveness. On the macro side, combinations of fiscal and monetary policy that yield high rates of inflation or high real interest rates worsen competitiveness of both exporters and import-competing producers by increasing the costs of production. High inflation likely leads to high rates of wage inflation. High real interest rates increase the cost of capital.

Policy combinations that yield an appreciating home currency worsen competitiveness through two channels. First, an appreciating home currency directly reduces the price competitiveness of exports and import-competing goods, although these direct effects can be mitigated somewhat by the indirect effect of the exchange rate movement on the cost of intermediates.

The direct effect of an appreciation also may be mitigated by the firm's decision to alter its markup to absorb some of the exchange rate change. To the extent that the currency movement is absorbed, international price competitiveness is retained. However, this decision is not costless. Reducing prices to maintain price competitiveness in the near term may require an offsetting strategy to recover profits at some future time. That is, the firm may be constrained to achieve a certain profit rate over some time horizon.[4] The firm's time horizon for profit maximization, the minimum profit rate, and expected movements for the exchange rate will partly determine its markup strategy. In this regard, the effects of macro policies on short-term exchange rate volatility and extended periods of exchange rate misalignment can have significant effects on pricing strategy and international competitiveness.

Policymakers are paying increased attention to how trade policies could improve international competitiveness. Of course there are many different policies, and only a few will be considered here. For import-competing industries, trade policies include leveling the playing field between domestic and imported goods by assessing duties on dumped or subsidized imports. For export industries, policies to improve foreign market access include reducing barriers to entry or voluntary import expansions. Using the model presented above, how would these policies work to improve competitiveness by either of the measures—relative prices or the trade balance?

Antidumping and countervailing duties raise the price of imports. Thus, by the relative price measure, competitiveness of domestic import–competing firms could improve as could external balance as consumers switch from imports to domestic substitutes. However, an important question is how the domestic import–competing firm reacts to less competition. In the context of the previous model of price determination, the import duty raises the import-competing price or reduces the number of competitors. A static, profit-maximizing domestic firm would increase its markup, which would offset some of the effects of the duty. On the other hand, if production technology is characterized by increasing returns to scale, the change in relative prices associated with the imposition of the duty could yield a greater demand for the domestic product, the domestic firm's costs could fall, and the firm could reduce the profit maximizing price. Thus production technology and the firm's strategic reaction to the duty are key to determining how international competitiveness might change with antidumping and countervailing duty policies.

Greater foreign market access might improve export price competitiveness through two mechanisms. Suppose that the restrictions on access to foreign markets can be proxied by increased costs of selling into the foreign market. In this case, the profit-maximizing export price would be higher in these markets than in other markets and exports to that market would be lower. By this definition, increased market access would reduce costs, the export price could fall, and exports rise. Moreover, price could fall if greater export volume reduces production costs. The trade balance measure of international competitiveness could improve. But note that the relative price of exports would fall, and by the terms of trade measure, international competitiveness would be worse, not improved.

Just as in the case of import duties, however, the strategic pricing behavior of exporters and the implementation of market access agreements would likely affect how the measures of international competitiveness change. Suppose that production economies of scale are a key reason for pursuing a market access deal. An important question is whether the exporters have exhausted all foreign markets besides the targeted market. If not, why is entry into this market necessary or large enough to change a firm's costs, price competitiveness, or profitability? Second, managed trade agreements can enforce a noncompetitive market structure in the destination market and keep prices and profits high for both domestic and foreign firms. In these cases, improved foreign market access may not lead to increased exports or improved international competitiveness.

Evidence on U.S. International Competitiveness

The above examples focused only on factors affecting the price competitiveness of U.S. firms. Similar models could be written for the foreign competing prices P^* and P_f. This suggests that a comparison between the United States and key foreign competitors of factors such as input costs, profit margins, and the exchange rate are important first places to look for sources of changes in U.S. international competitiveness.

Cost Differentials

Figure 3.5 shows measures of labor and capital costs for the United States, Japan, and Germany. The upper panel shows real average hourly compensation in manufacturing. The different patterns for these series across the countries is notable and suggests that labor compensation has not been a source of cost disadvantage for U.S. manufacturing firms. Real hourly compensation for Germany and Japan has risen steadily, while that for the United States has been flat.

The bottom panel shows a crude measure of the cost of capital.[5] Japanese firms may have enjoyed a substantial advantage over their U.S. and German counterparts until quite recently. In addition, U.S. and German firms apparently experienced significant increased costs of capital in the late 1970s while Japanese firms apparently did not. In the 1990s, the cost of capital to German firms apparently has been rising while that for U.S. firms has been falling.

Exchange Rate Movements

As noted earlier, the exchange rate has multiple avenues for affecting international competitiveness. With the time perspective shown in figure 3.3, the extent of the departure of the dollar exchange rate during the mid-1980s from its historical range is clear. While this movement was partly due to interest rate differentials during the period, it may partly represent a misalignment from the standpoint of more fundamental determinants of competitiveness, such as productivity differentials.

Moreover, the quarter to quarter volatility is notable. In recent years, while the real exchange value of the dollar has on balance depreciated, quarter-to-quarter movements of 10 percent have not been uncommon. This

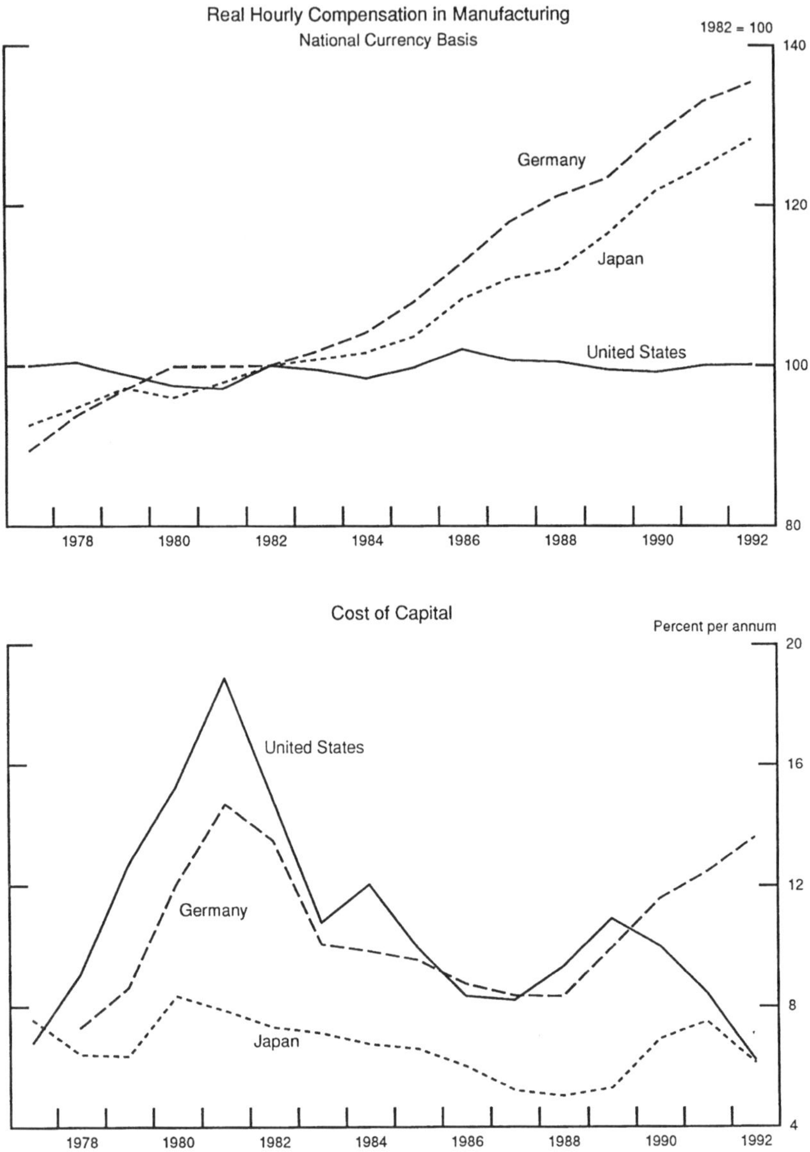

Fig. 3.5. Determinants of international competitiveness factor costs. (From Bureau of Labor Statistics, "International Comparison of Manufacturing Productivity and Unit Labor Cost Trends, 1992," table 13; IMF, various years, "International Financial Statistics," borrowing rates, line 60.)

volatility is notably greater than the plus-or-minus 2.25 percent exchange rate bands incorporated into the exchange rate mechanism of the European Monetary System.[6] Whether exchange rate volatility matters and over what frequency is controversial. Some argue that firms have available financial instruments to limit any deleterious effects of exchange rate volatility. These instruments do cost more as volatility rises, however.

Profit Margins

A third factor affecting international relative price competitiveness is the extent to which firms' profit margins change to offset cost and exchange rate movements. Figure 3.6 shows crude proxies for how profit margins of U.S. exporters and foreign importers may have changed over the last 20 years. Before 1982, the relationship between the proxies for costs (U.S. producer price index for exports and exchange-rate adjusted foreign CPI for imports) and price indices for exports and imports was rather similar. Between 1981 and 1985, profit margins appear to have narrowed for U.S. exporters, as costs continued to rise and export prices flattened. Foreign importers' margins appear to have widened as foreign cost inflation decelerated somewhat more quickly than did import prices. This is consistent with U.S. exporters attempting to offset some of the dollar appreciation by narrowing margins, while foreign importers were able to widen their margins even as they maintained dollar import prices in the United States. Since 1985, exporters have maintained tight margins, while importers' margins apparently have narrowed substantially, as cost inflation has apparently exceeded import price inflation.

While the behavior of profit margins is consistent with theories of pass-through and pricing-to-market, the profit margins of U.S. exporters appear less sensitive to exchange rate fluctuations than, apparently, are the profit margins of importers. There are several possible reasons for this behavior. For most U.S. producers, foreign market sales are a relatively small share of total production. Since maintaining separate foreign-currency price lists is not costless, less price and profit margin variability could result. Second, if the time horizon for U.S. firms is systematically shorter than for foreign firms, U.S. firms would tend to keep profit margins stable and allow prices to move with costs and the exchange rate.

Productivity

A final factor determining the evolution of international competitiveness is productivity differentials between the United States, its major competitors, and important new entrants to the international trading scene. Figure 3.7

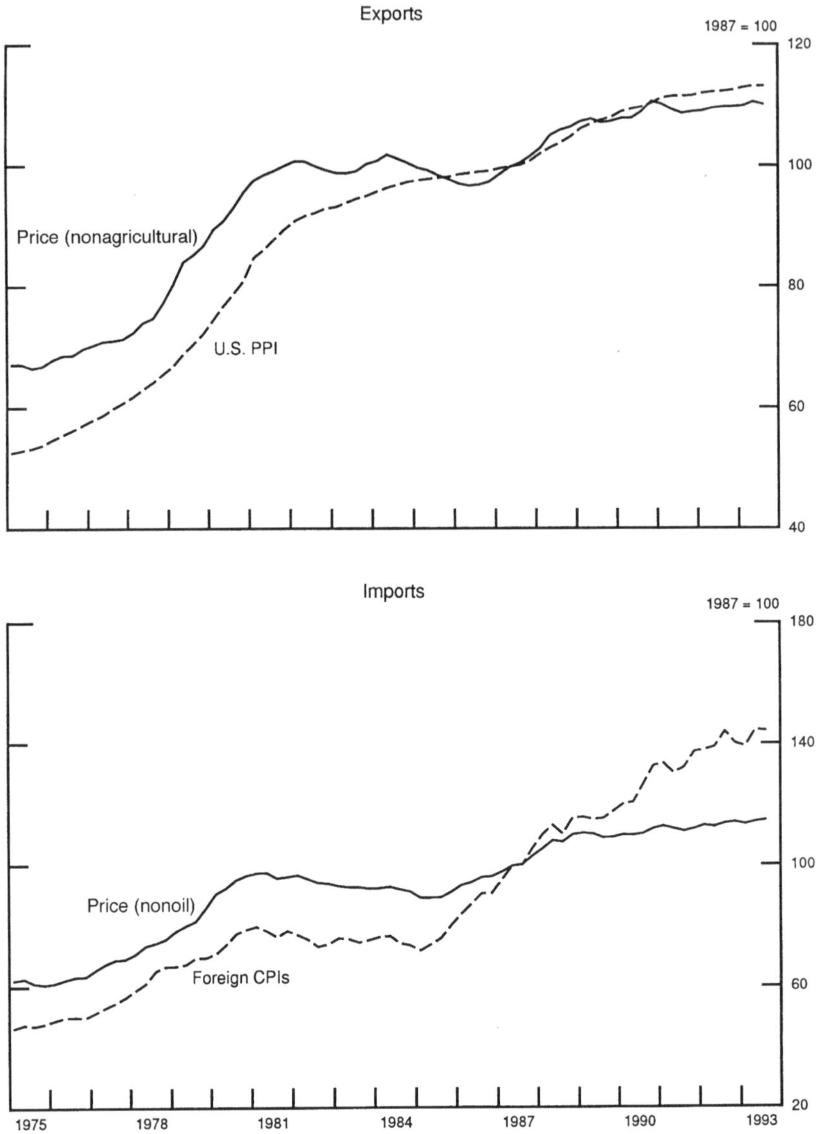

Fig. 3.6. Determinants of international competitiveness: Profit margins. (Author's calculations from unpublished data from the Department of Commerce and Bureau of Labor Statistics.)

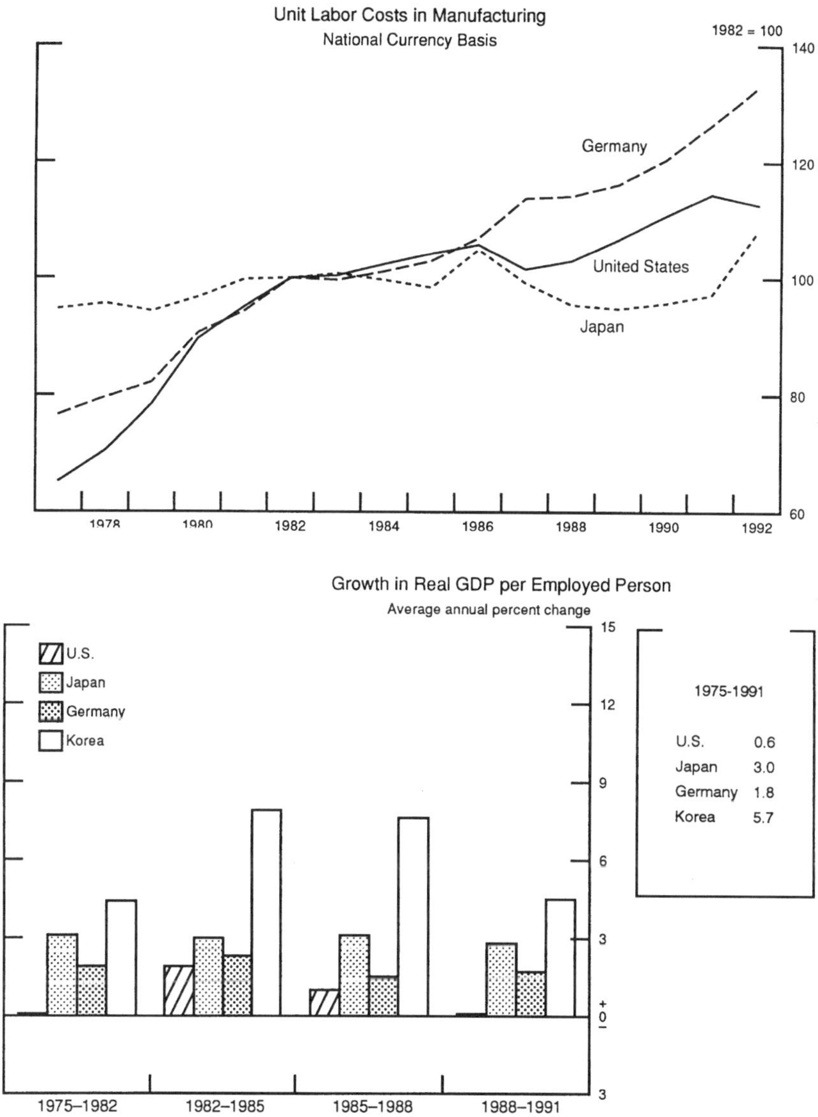

Fig. 3.7. Determinants of international competitiveness: Productivity growth. (From the Bureau of Labor Statistics, "International Comparison of Manufacturing Productivity and Unit Labor Cost Trends, 1992," table 9; Bureau of Labor Statistics, and unpublished data.)

shows two measures of productivity. The upper panel shows unit labor costs in manufacturing, which implicitly incorporates a measure of labor productivity. Whereas figure 3.5 indicated that labor compensation was flat in the United States, U.S. unit labor costs rose substantially through the early 1980s, flattened during the mid-1980s, and resumed a moderate increasing trend until 1992. This is consistent with slow labor productivity growth in the early 1980s and late 1980s to early 1990s. Unit labor costs in Germany have risen substantially over the whole period. In Japan, unit labor costs remained essentially flat until the sharp rise in 1992. Thus, compared with Germany, the productivity-adjusted cost competitiveness of U.S. labor should be substantially improved in recent years, while in comparison to Japan, the United States may be entering a period of increased competitiveness.

The bottom panel shows an economywide measure of productivity—real GDP per employed person—for the United States, Germany, Japan, and Korea. The poor average productivity performance for the United States over the 20-year period is notable, shown in the box at the right. For certain of the subperiods, U.S. productivity is close to zero. Note also the trend decline in the average annual growth rate of productivity for the three subperiods since the early 1980s.[7] In broad-brush terms, this trend has mirrored the deterioration in U.S. international competitiveness as measured by both the trade balance and the terms of trade, despite improvements in relative U.S. labor and capital costs.

Conclusions

The mathematical model previously presented suggested that economic policies affect international competitiveness through the building blocks of the firm's input costs and pricing strategy as well as by affecting the exchange rate. The evidence in the preceding section indicated that over the last twenty years, external balance has moved broadly in the expected way in response to changes in the exchange rate adjusted cost differentials between the United States and important trading partners. On the other hand, there has been a trend deterioration in external balance. Unfavorable productivity differentials may underlie this trend even as cost and exchange rate movements would point to an improvement in international competitiveness. From this evidence, what implications can be drawn for how best to enhance international competitiveness in the next decade?

First, looking at the period as a whole, movements in the real exchange rate have outweighed movements in any other determinant of international competitiveness. Since the exchange rate only reflects the underlying policy environment, macroeconomic policy management is key.

Second, input cost differentials are important determinants of international competitiveness. Macroeconomic policy management, as well as assuring competitive markets for inputs, is an important factor affecting input costs.

On top of the macroeconomic management, what role is there for managed trade policies? The historical decomposition of the U.S. partial trade balance suggested that most of the evolution of external balance was accounted for by income growth differentials, relative price changes, and exchange rate changes. For managed trade policies to have a systematic impact they must importantly contribute to one of these three factors. The theory of the firm allowed for managed trade policies to affect relative prices principally by changing the markup strategy. But the evidence suggests that movements in markups have not been more important for the international competitiveness of firms than have been movements in input costs or the exchange rate.

Finally, productivity growth is the key to long-term improved international competitiveness. Relatively slower productivity growth in the United States implies that the external gap probably cannot be closed without a deterioration in the standard of living. With relatively greater productivity growth, other countries will be able to produce more and more efficiently with their resources. While the United States will gain in terms of cheaper imports, it cannot continue to purchase more than it exports. Over a longer horizon, the exchange value of the dollar will systematically depreciate to offset the productivity differentials, to the detriment of U.S. well-being. To counter this scenario, the United States needs to generate relatively faster productivity growth. What would do this? A stable macroeconomic environment complemented by flexible and competitive markets.

NOTES

1. This decomposition is somewhat artificial in that the three factors—exchange rate, traded goods prices, and income growth—have interrelated effects on the trade balance.

2. The income terms are suppressed only for clear expositional focus on the role of the firm in relative price determination. Based on figure 3.4, understanding relative price movements (which include exchange rates) appears somewhat more important than understanding relative income growth differentials in explaining external balance.

3. Similar models of price determination could be presented for the foreign prices. Simplicity argues against doing so here, although we will assume such similar models later.

4. In the limit, of course, this hurdle is zero, otherwise the firm goes bankrupt and exits the industry.

5. The interest rate as a measure of the cost of capital is crude for a number of reasons. First, the real interest rate would be a better measure of the cost of debt. However, inflation differentials between these three countries are relatively small, and certainly smaller than the other errors in using either the real or nominal interest rate to measure the cost of capital. Second, the actual cost of capital to a firm depends significantly on its capital structure and the taxation regime of a country, both of which differ substantially across these three countries. Finally, open international capital markets mean that a national interest rate may bear no relation to the interest rate at which a national company might borrow, although again the degree of openness to international capital markets differs for firms in these three countries.

6. The EMS bands were widened to 15 percent in late 1993.

7. The most recent data for productivity show significant gains. It remains to be determined whether this improvement is long-term, or based on short-term fluctuations in labor input that are not indicative of a trend improvement in productivity.

PART 2

Competitiveness and Financial Markets

CHAPTER 4

The Anglo-Saxon Market for Corporate Control

The Financial System and International Competitiveness

Ajit Singh

This chapter was completed in 1995 at a time when there was a great deal of concern among academics as well as the business community about the international competitiveness of U.S. corporations. The MIT Commission on Industrial Productivity had found that the U.S. corporations had abandoned a whole range of industries to foreign competitors not because U.S. wages were too high but because of deficiencies of the U.S. financial system (Dertouzos, Lester, and Solow 1989). These deficiencies were thought to have also resulted in lower rates of investment particularly in R&D, in new product development, and in the upgrading of workers' skills compared with the corporations in Germany and Japan. With the U.S. economy enjoying at present (1998) one of its longest booms, these concerns have subsided. The catch-up by the U.S. and U.K. manufacturing industries with their German and Japanese counterparts in productivity growth has become more pronounced—early signs of this were noted in this chapter. The United States, however, still has a relatively large current account deficit that is likely to become more acute, partly as a consequence of large devaluations in Asian countries following their financial crisis. Further, as the U.S. economy slows down, the question of the international competitiveness of the U.S. corporations and the role of the U.S. financial systems are likely to command public attention again.[1]

The last decade has witnessed a growing debate on both sides of the Atlantic on the effectiveness of the stock market–based financial systems of the United States and the United Kingdom for promoting international competitiveness and industrial strength. The two countries share a broadly common framework of corporate law and possess the most advanced and complete stock markets in the world. As far as the corporate sector is concerned, in both countries, the financial system is more or less similar, being dominated by the stock market and a vigorously functioning market for corporate control. In principle, the latter is supposed to constitute an important additional mechanism by means of which the stock market can discipline firms and promote corporate efficiency. However, an increasing number of industrialists as well as academic economists argue that the Anglo-Saxon financial system is inferior to that of Japan and Germany and puts the United States and United Kingdom at a competitive disadvantage.

There is, of course, a long history of dissatisfaction in both countries with the financial system in general and the stock market in particular, but these criticisms have usually come from a minority of heterodox economists. Significantly, in the United States, the activities on the stock market have also often attracted popular suspicion and populist reaction. In the 1930s, the populist sentiment linked the stock market to the Great Depression. This led to the creation of the Securities and Exchange Commission and the passing of the Glass-Steagall Act and other measures to regulate the financial system. In the United Kingdom, academic critics have long argued that the financial system was in part responsible for the low rate of investment in the economy; specifically, it is suggested that the system facilitated investment abroad of domestic savings at the expense of investment at home to the detriment of the domestic economy. However, the financial establishment, as well as the mainstream of the economics profession, has traditionally maintained that the low rate of investment in the economy has not been due to the availability of finance but rather the lack of investment opportunities and that investment abroad has simply been due to the fact that the risk-adjusted rates of return on foreign investment have been greater than on home investment.[2]

However, today far-reaching criticisms of the Anglo-Saxon financial system come from the heart of the establishment itself. Thus, Michael Porter reported in the early 1990s on the results of a large research project on various aspects of the U.S. financial system, "the change in nature of competition and the increasing pressure of globalization make investment the most critical determinant of competitive advantage. . . . Yet the U.S. system of allocating investment capital both within and across companies is failing. This puts American companies at a serious disadvantage in global competition and ultimately threatens the long term growth of the U.S. economy."[3]

One cannot help noticing a certain irony in the fact that this skepticism about the virtues of the stock market is manifesting itself in the very citadels of these markets at a time when third world countries are falling over themselves to establish and to encourage such markets in order to promote economic development. Existing stock markets are being further developed or new ones being established from Kingston, Jamaica, to Ulan Bator in Outer Mongolia. More significantly, stock markets have been embraced by socialist China. In India, which now has more companies listed on the stock market (6,000 companies) than almost any other country, most leading cities already possess or are about to establish their own stock exchanges.[4]

The main object of this paper is to examine the role of the market for corporate control in the Anglo-Saxon financial system and to investigate its relationship to the question of international competitiveness. The causal links between the two—the market for corporate control and "international competitiveness"—are far from being obvious, even at a conceptual level, and some are quite controversial. An earlier related paper (Singh 1993b) analyzed various aspects of the market for corporate control in the United States and the United Kingdom. Specifically, the paper examined the diametrically opposite conclusions and analyses concerning the virtues of such a market, of industrial organization economists (see, e.g., Scherer 1988) on the one hand and of specialists in finance on the other (see, e.g., Jensen 1988, 1989). Much of that discussion bears directly on the present subject. The following sections will refer to and build on that analysis.

The Stock Market, the Market for Corporate Control, and International Competitiveness: The Causal Links

It is a striking fact that the rate of investment both in the United States and the United Kingdom has been appreciably less than in Japan and West Germany. On one estimate, manufacturing capacity in the United States increased by only 2.75 percent per annum in the 1980s, compared with 3.5 percent in the previous 15 years. Net investment fell from an average of 7 percent of GNP in the 1970s to 5 percent in the 1980s—the lowest rate in any of the big industrial economies other than the United Kingdom. In the case of the United Kingdom, it is estimated that if investment as a proportion of national expenditure had been the same as in West Germany, in almost every year in the 1970s and the 1980s, investment would have been 12 to 20 billion pounds higher (at 1989 prices) (Cosh, Hughes, and Singh 1990). Moreover, in the second half of the 1980s, which saw an unprecedented wave of corporate takeovers (Singh 1992), privately funded R&D spending in the United Kingdom fell to a lower level than in the other major economies.

In the early 1990s both the United States and the United Kingdom were running sizable current account and trade deficits, even at their high rates of unemployment. The causal link between this phenomenon and a low rate of investment is quite straightforward and generally accepted. In a world of imperfect nonprice competition in manufacturing products, economies with high rates of investment, other things being equal, are also likely to have higher rates of technical progress, new product development, and so on, which, in turn, will enhance their international competitiveness.

However, the causal connections between lower rates of overall or long-term investment and the stock market are much more controversial. The case of the critics of the Anglo-Saxon financial system runs along the following lines. At the simplest level, the argument is that the existence of a highly active market for corporate control, with its hostile takeovers and leveraged buyouts, obliges the U.S. and the U.K. managers to pay close attention to their earnings per share performances every quarter or every six months. This forces them to become "short-termist" in their outlook and to sacrifice long-term useful investments at the altar of short-term earnings.

At a more sophisticated level, the literature now contains several alternative and mutually reinforcing formulations of the short-termist position.

1. *Myopic markets.* This was the original argument of Keynes in chapter 12 of the *General Theory.* Keynes complained that "day-to-day fluctuations in the profits of existing investments, which are obviously of an ephemeral and non-significant character, tend to have an altogether excessive, and even absurd, influence on the market." In terms of this analysis, which clearly implicitly rejects the efficient market hypothesis of share price determination, the stock market induces short-termism even if there were no market for corporate control or fickle institutional fund managers. Neither corporate takeovers nor institutional investment were important elements in the functioning of the stock market at the time Keynes was writing. The Keynesian argument has been formalized in recent theoretical contributions.[5]

2. *Myopic managers.* In important contributions, Stein (1988, 1989) has established that even if the stock market were rational, it could be an optimal strategy for managers to be myopic and undertake short- rather than long-term value-maximizing projects. The argument rests on assuming imperfect and asymmetric information between shareholders and managers, the use of current earnings by managers to signal future prospects of the firm and the notion that there is "signal jamming," as it pays any manager not to provide accurate signaling information and to inflate the corporation's current earnings. The threat of takeovers is an integral part of this analysis as it implicitly assumes that, ceteris paribus, the lower the share price of a firm, the greater its likelihood of being taken over.

At a much simpler level, managerial myopia can also arise from inappropriate incentive structures for the managers—for example, one that implicitly or explicitly links managerial compensation to short-term performance measures.

3. *The role of the institutional investors.* Increasingly, corporate share ownership in the United States and the United Kingdom is being concentrated in financial institutions, particularly the pension funds and insurance companies. In the United Kingdom, these institutions own almost three-quarters of the shares of the companies quoted on the stock market (Cosh, Hughes, Lee, and Singh 1989). In the United States the corresponding figure is nearly 60 percent (Ghilarducci 1992). In principle, with typical long-term liabilities, institutional share ownership should lead to long term value maximization and to "patient" capital. In practice, it is suggested that because of the particular structural features of institutional fund management the opposite situation prevails. Fund management is a highly competitive industry, and increasingly the performance of fund managers themselves is assessed on the basis of short-term results. This leads to high share turnover and acceptance of takeover bids on the basis of short-term financial gain rather than long-term industrial logic.[6] The latter behavioral pattern also involves a phenomenon of the herd instinct, which in turn may be attributed to the concept of asymmetric payoff. It is pointed out that there are sound reasons for fund managers to display the herd instinct; if a fund manager who does not follow the herd turns out to be wrong in his investment policies when the herd is right, he or she may be subject to severe penalties, such as dismissal. However, if the herd is wrong and the fund manager is right, his reward is not commensurate. Faced with the prospect of an immediate stock market gain from a takeover situation, most fund managers are likely to follow the herd and accept it rather than not.

4. *Contrast with the German and Japanese Financial Systems.* A very important part of the argument of the critics of the Anglo-Saxon financial system rests on its contrast with the financial systems in Germany and Japan. In the latter two countries, for a number of legal and institutional reasons, the stock market has relatively little influence on the performance and behavior of industrial companies. The following differences between these financial systems are particularly important in the context of the present discussion.

First, in sharp contrast to the situation in the United States and the United Kingdom, there are hardly any hostile takeovers in Japan or West Germany. The main reason for this phenomenon is that compared with Anglo-Saxon countries, the Japanese or the German shareholders have far less effective power than other stakeholders in the corporation, for example, the corporation's workers, suppliers, and customers. In West Germany, the right of workers in large companies to be consulted over important investment and employment decisions is institutionalized through the 1976 Co-determination Act and by employee representation on the corporate supervisory boards. This system of corporate governance is a major barrier to hostile takeovers in Germany, which is reinforced by other features of the system outlined below. Similarly in Japan, in addition to other aspects, Odagiri and Hase (1989) suggest that the system of lifetime employment in large companies plays an important role in deterring involuntary takeover bids.

The second important feature of the German and Japanese systems is the concentration of corporate share ownership in a relatively small number of "safe" hands. In Japan, for example, in a typical large corporation, almost three-quarters of the shares are likely to be held by its suppliers, customers, and banks. There are similar complex cross-holdings of shares in Germany.

Thirdly, in relation to Germany, Hart (1992) notes that there are only 545 large German companies with a stock exchange quotation (plus 94 in the unlisted securities market). There is, however, active trading in shares of a small handful of these companies—about 30 or so. These are probably the only large companies where majorities of voting shares are not controlled by a parent company, families, or other institutions (particularly banks) closely linked to incumbent management. In the case of Japan, Dore (1985) argues that not only is the stock market viewed with suspicion by the general public, it also has a rather inferior social status. It is the real wealth-creating corporate sector or the government that attracts the best talent rather than the stock market. For all these and other reasons, it is argued that the bank-based German and Japanese financial systems are more conducive to long-term investment not only in plant and equipment but more importantly in training and in firm-specific and often intangible human capital.[7] In the case of the United States, Shleifer and Summers (1988) had suggested that the microeconomic private efficiency gains from takeovers, even to the extent that they exist (see below), may greatly overstate the social benefit. Increased post-takeover profitability may not represent a genuine improvement in social efficiency, as it may simply be a transfer of resources from one group of stakeholders (for example, employees) in the firm to another (the shareholders). Shleifer and Summers went on to suggest that such transfers, which takeovers facilitate, for various institutional reasons may be socially harmful in that they involve a breach of trust and the breaking of implicit contracts between managers and workers. It is on the basis of trust and these associated implicit contracts that workers undertake firm-specific training that leads to greater productivity and thus benefits both the firm and the economy. However, as Jenkinson and Mayer (1992) rightly note, it is not so much that hostile takeovers lead to a breach of implicit contracts, but that under this system such contracts are unlikely to emerge at all.

The Anglo-Saxon Financial System: The Case for the Defense

Against the above indictment, the following broad arguments can be made in defense of the system.

1. First, there is a prior argument. Is there an Anglo-Saxon competitive problem at all that needs to be addressed? As table 4.1 shows during the period 1973 to 1990, United States manufacturing production increased at a faster rate than in Germany or in the EU as a whole, excluding Germany. True, Japanese manufacturing production grew more rapidly than that of the United States in the post-1973 period, but even here, compared with the previous experience (1950–73), the gap has narrowed very considerably. Similarly, the rate of growth of manufacturing productivity in the United States is now more or less the same as in the West European countries; there is also a much smaller handicap with respect to Japan compared with before.

The comparative performance of U.K. manufacturing has also greatly improved in the 1980s compared with its position in the 1960s and 1970s. O'Mahony (1992) and Prais (1990) report that in the early 1980s the level of manufacturing productivity in West Germany was 33 percent greater than that in the United Kingdom—the gap being considerably larger in, say, the mid-1960s. However, by 1987, the difference in the levels of manufacturing productivity in the two countries had been reduced to 22 percent. Today in the light of the problems of German reunification, the German advantage will be even smaller.

2. Hatsopoulos, Krugman, and Summers (1988) forcefully argue that the problems of low U.S. investment and short time horizons of U.S. corporations stem essentially from the fact that the cost of capital in the United States is higher than in Japan and in other competitor countries. However, they ascribed this phenomenon entirely to macroeconomic causes, specifically the much lower rate of U.S. savings compared with, for example, Japan. Between 1981 and 1985, the U.S. national savings as a proportion of GNP averaged only 3.2 percent; the corresponding figure for Japan for that period was 10.6 percent.

On this argument, then, the problems of short-termism and low rate of investment in the United States have little to do with myopic markets, or myopic managers, or with impatient institutional investors and other failings of the U.S. financial system. Rather, Hatsopoulos, et al. suggest:

> The short-term focus of American managers . . . is a rational response to their market situation. If anything, the evidence from takeovers and market valuations suggests that American managers take a longer view

TABLE 4.1. Manufacturing Performance in U.S., Japan, and EEC, 1950–97 (annual growth rates)

	Output			Output per Worker		
	1966/1950	1973/1966	1990/1973	1966/1950	1973/1966	1990/1973
United States	4.5	3.2	2.6	3.0	2.6	2.5
Japan	14.8	14.4	3.1	8.7	11.1	5.3
Germany	7.8	5.3	1.6	4.2	5.1	2.4
EEC, excluding Germany	5.0	6.0	1.8	3.3	5.9	2.8

Source: Adapted from Rowthorn 1992, table 3; OECD National Accounts 1995 (OECD: Paris).

than is in the best interests of their stockholders. The appropriate strategy for lengthening business planning horizons and encouraging long-term investments is not to criticize corporate cultures but to change the market incentives that shape them. . . . It is important to emphasize the role of the macroeconomic environment, of which the cost of capital is a key feature, in our competitive performance. (303)

3. Michael Jensen, in a number of influential contributions (see, e.g., Jensen 1988, 1989), has robustly defended the merits of the market for corporate control in the United States. He writes: "The market for corporate control is creating large benefits for shareholders and for the economy as a whole by loosening control over vast amounts of resources and enabling them to move more quickly to their highest-valued use. This is a healthy market in operation, on both the takeover side and the divestiture side, and it is playing an important role in helping the American economy adjust to major changes in competition and regulation of the past decade" (1988, 23).

Specifically, Jensen suggests that gains from takeovers in the United States amount historically to 8 percent of the total value of both companies. In his view, these value gains represent "gains to economic efficiency, not redistribution between various parties." He commends the activities of takeover specialists such as T. Boon Pickens as benefiting shareholders and, hence, the society. He rejects the myopic markets hypothesis on a number of different analytical and empirical grounds. He appeals to an empirical study by the Office of the Chief Economist of the SEC (carried out in 1985) to suggest that takeovers do not discourage R&D spending.

For good measure, Jensen (1989) asserts that unless the Japanese establish for themselves a market for corporate control of the kind that exists in the United States, the Japanese corporations will become bureaucratic and inefficient and lose their competitive edge in the world markets.

The Case for the Defense: A Critical Analysis

We shall now briefly analyze each of the preceding points in turn.

1. First, on the question of improved Anglo-Saxon comparative industrial performance in the 1980s, two points may be made. First, as noted earlier, both countries continued to run large current and trade deficits in the early 1990s, despite the fact that they have high rates of unemployment. In the U.K. case, there is little evidence of any structural change during the last decade either in the country's relatively high long-term propensity to import manufactures, or in the world income elasticity of demand for U.K. manufactured exports. The problems of U.K. deindustrialization can be traced essentially to these adverse import and export elasticities. The undoubted improvement in the comparative record of growth of manufacturing productivity in the United Kingdom during the last decade has not helped to alter these critical structural variables. As a consequence it is estimated that the country will only be able to maintain a current account balance at high rates of unemployment.[8]

Second, in relation to the U.K. economy, it has been observed that its *comparative* performance improves when the world economy is growing slowly and deteriorates when the latter is growing fast. This stylized fact is interpreted to suggest that the U.K. economy is usually supply-constrained while those of competitor countries tend to be demand-constrained. Thus, the comparative improvement in the U.K. industrial record in the 1980s reflects more a deterioration in the performance of other countries than a genuine betterment of the country's industrial economy (Michie 1992).[9]

2. On the issue of short-termism being primarily a macroeconomic phenomenon, caused by high cost of capital arising from a lower U.S. savings rate, the following points are in order. First, even if one were to accept the proposition that high cost of capital leads to short-termism, it does not imply that it is not caused by other factors as well. Myopic managers or myopic markets, and so on, may reinforce the effects of the cost of capital.

Secondly, turning to the cost of capital itself, is it really entirely a macroeconomic phenomenon? Surely, it is influenced by microeconomic as well as institutional factors. For example, for a given savings rate in an economy, the cost of capital to the corporate sector is likely to be higher the more variable and unstable the economic performance. This line of reasoning provides an important link between the micro- and macroeconomic variables. For, if as a result of myopic stock markets, managers do not invest enough in training or other forms of long-term investment, it will make the economy less internationally competitive and therefore more subject to stop-go syndrome and more unstable at the macroeconomic level.

Similarly, it has been rightly emphasized that, even with the same savings rate, the cost of capital in the corporate sector in Japan and Germany would be lower than in the Anglo-Saxon countries because of the particular institutional arrangements for the provision of corporate finance in the former countries. As the banks in Germany and Japan have close relationships with the corporations, and they closely monitor corporate performance, the problem of asymmetric information is greatly attenuated. Other things being equal, this should lead to a lower cost of outside finance (Singh and Hamid 1992).

3. Turning to Michael Jensen's favorable analysis of the role of takeovers, Singh 1993b contained a detailed examination of his various propositions. Very briefly, in principle, the takeover mechanism can promote economic efficiency through two distinct channels: (a) the *threat* of takeovers, operating via the market for corporate control, may discipline firms that are not making the best use of the resources under their command; (b) even if the firms were operating efficiently, the *actual* takeovers and mergers themselves may lead to a reorganization and restructuring of an economy's resources and thereby enhance their social value. Research during the last two decades has shown that there are a number of important reasons (imperfections in the capital market, the free-rider problems, transaction costs, etc.) why neither of these channels may work effectively in the real world. More importantly, empirical studies show that contrary to Jensen and the folklore of capitalism, takeovers do not simply punish the inefficient and the unprofitable and select for survival companies that best enhance shareholder wealth. These studies unequivocally conclude that although takeover selection does take place to a limited degree on the basis of profitability or stock market valuation, it also importantly does so in terms of size. Thus in the market for corporate control a large unprofitable corporation has a much higher chance of survival than a small relatively much more profitable company. As a consequence, the *threat* of takeover, instead of forcing firms to improve their profitability, may in fact encourage them to further increase their size. This perverse result of the takeover mechanism may be compounded by the takeover mechanism itself since it may enable relatively inefficient large firms to grow larger still by taking over more efficient smaller firms.[10]

Further, for detailed reasons given in Singh 1993b, Jensen's assertion, that the substantial bid premiums paid for taken-over firms constitute "gains in economic efficiency," simply cannot be accepted. His view rests on the assumption that share prices in the real world are always efficient in the fundamental valuation sense of Tobin. If this assumption is not accepted—and there are important analytical as well as empirical grounds for not doing so—the big premiums can be interpreted in a rather different way. Charkham

(1989) has put forward a dual valuation view of stock market pricing of the takeover victims, which provides a more plausible interpretation of the observed facts. In this conception there is a normal day-to-day valuation of the company's shares, based on its expected earnings and reflecting valuation at the margin. However, there is a higher alternative valuation when the company is put into play and is subject to a takeover bid; this reflects the price to be paid for buying the intramarginal shareholders in order to gain control of the company (Shleifer 1986). If the raiders are empire-building managers they may be willing to offer a large premium for control, especially as they are paying with money that is not their own.

Conclusion and Proposals for Reform

1. It has been argued here that the market for corporate control today provides an important link in the chain of causation connecting the Anglo-Saxon financial system with short-termist corporate culture, low investment in long-term projects (particularly training and firm-specific human capital), and hence diminished international competitiveness. The supposed benefits that orthodox economic analysis ascribes to the market for corporate control do not materialize in practice. Evidence indicates that the operations of the market not only do not enhance efficiency but the results of the takeover process may be perverse in a number of ways.

As a disciplinary device, the takeover mechanism is shown to have serious shortcomings, particularly as far as the very large firms are concerned. Further, with respect to the workings of the takeover selection process, it is not at all clear that the managements that are selected for survival are necessarily the better managements from the point of view of creating real wealth, rather than simply being more skilled at financial engineering.

2. Available research does not allow us to assess the relative quantitative significance of short-termism and that of a whole host of other factors (e.g., the English class system, labor management relations) that may be responsible for the international competitive disadvantage of the Anglo-Saxon countries. However, what is quite clear from a number of surveys carried out both in the United States and the United Kingdom is that corporate leaders in both countries believe that the Anglo-Saxon financial system puts them at a competitive disadvantage with their counterparts in Germany and Japan.[11] Hart (1992, 2) notes in relation to the United Kingdom "that even if there is no conclusive evidence that City financial institutions are biased against companies with long-term research programmes, industrialists think that they are and opt for short-term results to protect their share price. This belief may be wrong, but it is very powerful nevertheless."

3. Although in the city-industry divide each side blames the other for short-termism, I agree with Michael Porter that the problem is a systemic one. The Anglo-Saxon corporate executives and institutional fund managers are both faced with a market environment and an incentive system that leave them with little room to maneuver, except to give primary importance to short-term financial gain. Instead of a system that places a premium on launching and resisting takeover bids and on skills in financial engineering, the Japanese and German financial systems provide the corporate executives with a rather different set of incentives. The latter favor organic growth and corporate investment that are directly beneficial to the economy.

4. There is a whole plethora of proposals both in the United States and the United Kingdom for reforming the system. These are usually piecemeal proposals directed at one aspect or the other of the system—for example, the regulation of takeovers, the structure of corporate law or the fiscal status of the shareholders, and so forth. However, they do not view the system as a whole. Michael Porter's recent work does provide recommendations that consciously address the systemic problems of the U.S. financial and industrial systems.

Since it is not practical to reform everything at the same time, the first important issue in reforming an interdependent and complex system is to identify the causal link and the area where the reform effort should be concentrated. The centerpiece of Porter's proposals is his "long-term equity investment incentive." Broadly, this entails an incentive for holding shares in manufacturing corporations for a minimum of a five-year holding period, with greater incentives provided for even longer holdings. The incentives would be applied prospectively to new investments. Porter (1992b, 152) believes that this is the "single most powerful tool for changing the U.S. system." He argues that "if investors have a reason to hold shares for five years or more, the whole approach for investing in America would change." The market players would then be interested in the long term prospects of the company rather than in the "current game of predicting near-term share price movements."

5. My own proposal for reform would concentrate on a different part of the system—the nature and extent of property rights of shareholders of large corporations. In the Anglo-Saxon system of corporate law, the property rights for buying and selling corporations lie entirely with the shareholders of the corporation. Is it desirable or economically efficient that the right to dispose of the whole corporation should solely be the prerogative of the mostly absentee shareholders of the company without reference to any of the other interested parties (e.g., the workers)?

Whether or not it is otherwise desirable, I believe it will be economically efficient for serious restrictions to be placed on the existing property rights of shareholders in the large U.S. corporations. Specifically I would propose that long-term employees of the corporation, as well as the broader community where the corporation is located, should have a legally recognized voice—but not necessarily a decisive one—in major corporate decisions such as a takeover of the corporation as a whole or a divesture of a large part of its assets. This will not only help reduce the incidence of takeovers for short-term considerations but, depending on the institutional form in which this proposal is implemented, it should also lead to better labor-management relations.

On the face of it, changing property rights and the basic structure of corporate law may appear to be a radical proposal, which may be thought to undermine the capitalist system. But in fact we find that capitalists outside the United States and the United Kingdom have no difficulty in accepting some restrictions on property rights to meet a broader social purpose. The German system of corporate governance, with its two-tier boards, entails important restrictions on property rights of the owners. Yet the German corporate executives seem quite content with it, and the capitalist system works there very well. So one way of implementing the proposal I have outlined would be to institute in the United States and the United Kingdom a suitable variant of the German system of corporate governance—one in which half of the corporation's upper-level supervisory board consists of representatives of the employees.

The two-tier board is of course not the only way in which workers can be given a statutory right to be consulted on important corporate decisions.[12] These ideas will be elaborated further in a forthcoming paper. Finally, I note that a proposal of the kind I have outlined above was put forward by Keynes as long ago as 1927 as a part of the manifesto for the U.K. Liberal Party.

NOTES

1. Please note the text of this chapter has not been changed since its completion in 1995, although some of the notes have been updated.

2. See, for example, Feinstein 1992; Wilson 1978; Ball 1991.

3. Porter 1992a, 65. This paper reports the findings of a large research project sponsored by the Harvard Business School and the Council on Competitiveness, a project that included 18 research papers by 25 academic experts.

4. For a discussion of the role of the stock market in socialist countries and in economic development, see Singh 1989, 1990, 1993a. For further information on the Indian stock market, see Singh 1996, 1998a.

5. See, for example, Shleifer and Vishny 1990. See also Shleifer and Summers (1990) and the papers by Scarf and Shiller, respectively, in Coffee 1987.

6. For a fuller discussion of this argument, see Cosh, Hughes, and Singh 1990.

7. With the German and particularly the Japanese economies not performing well at all in the second half of the 1990s, it is the corporate and financial systems of these countries that are being held responsible for their crises. It is, however, important to note that as far as international competitiveness is concerned, the two countries are still running sizable current account surpluses, and in a world of free trade (without so-called voluntary restraint agreements and threats of antidumping measures), the Japanese surplus would certainly be even larger. The crises in these countries are actually attributable more to macroeconomic factors than to the weakness of their financial and corporate structures. See the section below, "The Case for the Defense: A Critical Analysis," and Singh 1998a, 1998b.

8. There is a large literature on this subject. See Singh 1977, 1987, 1989; Coutts, Godley and Rowthorn 1986; Coutts and Godley 1990. For a different perspective, see Ball 1990; for a response to the issues raised in that paper, see Singh 1989. The more recent literature on this subject is contained in the Controversy on the question of deindustrialization in the *Economic Journal* (January 1996). See also the introductory chapter by Howes and Singh in this volume.

9. For more recent discussion of this subject, see the Controversy in the *Economic Journal* (January 1996); in particular, Kitson and Michie 1996 and Crafts 1996.

10. For reviews of the literature, see Singh 1992; Hughes 1995.

11. See Hart 1992 for the United Kingdom and Porter 1992a for the United States.

12. See Blair 1995.

REFERENCES

Ball, James. 1990. Manufacturing Industry, Economic Growth and the Balance of Payments. In *Money, Trade and Payments,* ed. D. Cobham, R. Harrington, and G. Zis. Manchester: Manchester University Press.

———. 1991. Short Termism—Myth or Reality? *National Westminster Bank Quarterly Review* (August): 20–30.

Blair, Margaret M. 1995. *Ownership and Control: Rethinking Corporate Governance for the Twenty-First Century.* Washington, D.C.: Brookings Institution.

Charkham, J. 1989. Corporate Governance and the Market for Control of Companies. Bank of England Panel Paper, no. 25.

Coffee, John, et al. 1987. *Knights, Raiders, and Targets.* Oxford: Oxford University Press.

Cosh, A. D., A. Hughes, K. Lee, and A. Singh. 1989. Institutional Investment, Mergers and the Market for Corporate Control. *International Journal of Industrial Organisation* 7, no. 1: 73–100.

Cosh, A. D., A. Hughes, and A. Singh. 1980. The Causes and Effects of Takeovers in the UK: An Empirical Investigation for the Late 1960s at the Micro-Economic Level. *In the Determinants and Effects of Mergers,* ed. D. C. Mueller. Cambridge, Mass.: Oelschlager, Gunn and Hain.

———. 1990. Takeovers and Short Termism: Analytical and Policy Issues in the UK Economy. In *Takeovers and Short Termism in the UK,* Industrial Policy Paper No. 3, Institute for Public Policy Research, London.

Coutts, K. J., and W. Godley. 1990. *Britain's Economic Problems and the Policies in the 1990s.* Economic Study no. 6, Institute for Public Policy Research, London.

Coutts, K. J., W. Godley, and R. E. Rowthorn. 1986. The British Economy: Recent History and Medium Term Prospects. A Cambridge Bulletin on the Thatcher Experiment, Cambridge (March).
Crafts, Nicholas.1996. Deindustrialisation and Economic Growth. *Economic Journal* 106 (January): 172–183.
Dertouzos, Michael L., Richard K. Lester, and Robert M. Solow. 1989. *Made in America: Regaining the Productive Edge.* Cambridge: MIT Press.
Dore, Ronald. 1985. Financial Structures and the Long Term View. *Policy Studies* 6 (July): 10–29.
Feinstein, Charles H. 1992. Capital Markets and Capitalism in Britain and Continental Europe before 1914. *The New Palgrave Dictionary of Money and Finance* 1:305–11. London: Macmillan; New York: The Stockton Press.
Ghilarducci, Teresa. 1992. Are Pension Funds Patient Investors?: Management Matters. Unpublished paper, Department of Economics, University of Notre Dame, Notre Dame, Ind.
Hart, Peter E. 1992. Corporate Governance in Britain and Germany. National Institute for Economic and Social Research, Discussion Paper no. 31.
Hatsopoulos, George N., Paul Krugman, and Lawrence Summers. 1988. U.S. Competitiveness: Beyond the Trade Deficit. *Science* 421 (July 15): 299–307.
Hughes, A. 1995. Mergers and Economic Performance in the U.K.: A Survey of Empirical Evidence 1950–1990. In *European Mergers and Merger Policy,* (2d ed.,) J. Fairburn and J. A. Kay. Oxford: Oxford University Press.
Jenkinson, Tim and Colin Mayer. 1992. The Assessment: Corporate Governance and Corporate Control. *Oxford Review of Economic Policy* 8, no. 3: 1–10.
Jensen, Michael C. 1988. Takeovers: Their Causes and Consequences. *Journal of Economic Perspectives* 3, no. 2: 21–48.
———. 1989. Eclipse of the Public Corporation." *Harvard Business Review* 67, no. 5: 61–74.
Kitson, Michael, and Jonathan Michie. 1996. Britain's Industrial Performance since 1960: Underinvestment and Relative Decline. *Economic Journal* 106 (January): 196–212.
Meeks, G. 1977. *Disappointing Marriage: A Study of the Gains from Mergers.* Cambridge: Cambridge University Press.
Michie, Jonathan, ed. 1992. *The Economic Legacy, 1979–1992.* Pullman, Wash.: Academic Press.
Mueller, D. C., ed. 1980. *The Determinants and Effects of Mergers.* Cambridge, Mass.: Oelscgalger, Gunn and Hain.
Odagiri, H., and T. Hase. 1989. Are Mergers and Acquisition Going to Be Popular in Japan Too? An Empirical Study. *International Journal of Industrial Organisation* 7, no. 1: 49–72.
O'Mahony, M. 1992. Productivity Levels in British and German Manufacturing Industry. *National Institute Economic Review,* no. 139 (February): 46–63.
Porter, Michael E. 1992a. Capital Disadvantage: America's Failing Capital Investment System. *Harvard Business Review* 70 (September–October): 65–82.
———. 1992b. Michael E. Porter Replies. *Harvard Business Review* 70 (November–December): 151–154.
Prais, S. J., ed. 1990. *Productivity, Education and Training.* London: NIESR.

Ravenscraft, D. J., and F. M. Scherer. 1987. *Mergers, Sell Offs and Economic Efficiency.* Washington, D.C.: The Brookings Institution.

Rowthorn, Robert E. 1992. Productivity and American Leadership. *Review of Income and Wealth,* series 38, no. 4 (December): 475–96.

Scherer, F. M. 1988. Corporate Takeovers: The Efficiency Argument. *Journal of Economic Perspectives* 2, no. 1: 69–82.

Schwarz, S. 1982. Factors Affecting the Probability of Being Acquired. *Economic Journal* 92 (June): 391–98.

Shleifer, Andrei. 1986. Do Demand Curves for Stock Slope Down? *Journal of Finance,* no. 41: 579–90.

Shleifer, Andrei, and Lawrence H. Summers. 1988. Breach of Trust in Hostile Takeovers. In *Corporate Takeovers: Causes and Consequences,* ed. A. J. Auerbach. Chicago and London: University of Chicago Press.

———. 1990. The Noise Trader Approach to Finance. *Journal of Economic Perspectives* 4, no. 2: 19–33.

Shleifer, Andrei, and Robert W. Vishny. 1990. Equilibrium Short Horizons of Investors and Firms. *American Economic Review, Papers and Proceedings* 80, no. 2: 148–53.

Singh, Ajit. 1971. *Takeovers: Their Relevance to the Stock Market and the Theory of the Firm.* Cambridge: Cambridge University Press.

———. 1975. Takeovers, Economic 'Natural Selection', and the Theory of the Firm: Evidence from the Post-war UK Experience. *Economic Journal* 85:497–515.

———. 1977. UK Industry and the World Economy: A Case of De-Industrialisation? *Cambridge Journal of Economics* 1 no. 2 (June): 113–36.

———. 1987. Manufacturing and De-Industrialisation, *The New Palgrave: A Dictionary of Economics.* London: Macmillan; New York: Stockton.

———. 1989. Third World Competition and the De-Industrialisation in Advanced Countries. *Cambridge Journal of Economics* 13 no. 1 (March): 103–20.

———. 1990. The Institution of a Stock-Market in a Socialist Economy: Notes on the Chinese Economic Reform Programme. In *The Chinese Economy and Its Future: Achievements and Problems of Post-Mao Reforms,* ed. Don Fureng and Peter Nolan. Cambridge: Polity Press.

———. 1992. Corporate Takeovers. In *The New Palgrave Dictionary of Money and Finance,* ed. John Eatwell, Murray Milgate, and Peter Newman. London: Macmillan; New York: Stockton.

———. 1993a. Regulation of Mergers in the US and the UK: A New Agenda. In *Industrial Economic Regulation: A Framework and an Exploration,* ed. Roger Sugden. London: Routledge.

———. 1993b. The Stock Market and Economic Development: Should Developing Countries Encourage Stock Markets? *UNCTAD Review 1993,* no. 4: 1–28.

———. 1996. The Stock Market, the Financing of Corporate Growth and Indian Industrial Development. *Journal of International Finance* 4, no. 2: 1–17.

———. 1998a. Liberalisation, the Stock Market and the Market for Corporate Control: A Bridge Too Far for the Indian Economy? In *India's Economic Reform and Development,* ed. I. J. Ahulwalia and I. M. D. Little. New Delhi: Oxford University Press.

———. 1998b. Financial Crisis in East Asia: The End of the Asian Model? ILO Discussion Paper no. 24, Geneva.

Singh, Ajit, and J. Hamid. 1992. *Corporate Financial Structures in Developing Countries.* Washington, D.C.: IFC/World Bank.

Stein, Jeremy C. 1988. Takeover Threats and Managerial Myopia. *Journal of Political Economy* 96:61–80.

———. 1989. Efficient Stock Markets, Inefficient Firms: A Model of Myopic Corporate Behavior. *Quarterly Journal of Economics* 104, no. 4: 655–69.

Warshawsky, M. J. 1987. Determinants of Corporate Merger Activity: A Review of the Literature. Staff Study no. 152, April, Board of Governors of the Federal Reserve System, Washington, D.C, summarized in *Federal Reserve Bulletin* 73, no. 4: 270–71.

Wilson, Harold. 1978. *Evidence on the Financing of Industry and Trade,* vols. 1–8, Committee to Review the Functioning of Financial Institutions, HMSO.

CHAPTER 5

American Corporate Finance

From Organizational to Market Control

William Lazonick and Mary O'Sullivan

Industrial Development

Over time, the innovation process and the learning process that is its social substance have become increasingly collective and cumulative, and hence organizational (O'Sullivan 1996; Lazonick and O'Sullivan 1996b). Our perspective on industrial development identifies *organizational integration* and *financial commitment* as the social conditions that permit collective and cumulative learning to take place. Organizational integration describes the social relations that provide participants in a complex division of labor with the abilities and incentives to integrate their capabilities and efforts within organizations so that they can potentially generate organizational learning. Financial commitment describes the social relations that are the basis for the ongoing access of a business organization to the financial resources required to sustain the development and utilization of productive resources.

In combination, organizational integration and financial commitment provide social foundations for innovative business enterprise. In terms of inputs into the production process, organizational integration supplies skills, and financial commitment supplies money. In contributing to the innovation process, however, these inputs are not commodities. They reflect the social relations to the business organization of people who supply skills and money.

The generation of innovation through organizational learning is inherently uncertain. The investment strategy that will result in a higher quality, lower cost product cannot be known in advance. What one learns as the innovation process evolves changes how one conceives of the problems to be addressed, the possibilities for their solution, and therefore the appropriate investment strategy for continued learning. Learning may make possible the attainment of ends that were previously considered impossible, and a restructuring of the learning process may be required to achieve these ends (O'Sullivan 1996).

In allocating resources to organizational learning, strategic decision makers must know what the learning process is generating if they are to take account of opportunities for, and threats to, innovative success that learning reveals. In the innovation process, strategic decisions shape the direction and structure of learning, and the knowledge continually generated through learning can inform strategy. For such a dynamic interaction of strategy and learning to occur, strategists must be integrated into the network of social relations that underlies the generation and transmission of organizational learning. If, instead, strategic control is exercised by decision makers who are segmented from the process of organizational learning, the ongoing innovative success of the enterprise will be jeopardized because decision makers will have neither the abilities nor incentives to promote that success.

The financial commitment required to sustain organizational learning on an ongoing basis means that strategic decision makers must exercise organizational control over the revenues of the enterprise. In contrast, those who exercise market control over enterprise revenues will favor financial liquidity, not financial commitment (O'Sullivan 1996; Lazonick 1992). Placed in the perspective of the process of industrial development, the distinction between organizational control and market control has great practical importance for current debates over corporate governance and corporate employment. In all of the advanced economies, different groups are currently contending for control over the allocation of vast corporate revenues. In making their arguments for creating value for shareholders, the proponents of market control rely on a theory that touts the efficacy of the market mechanism for the allocation of resources to their best alternative uses. But their theory offers no explanation of how the best alternative uses come into being or change over time.

Currently, the United States (along with Britain) is considered a bastion of market control.[1] Yet, as we shall show, market control over the allocation of corporate resources is a relatively new phenomenon in U.S. industrial history. Until the 1980s organizational control dominated in the allocation of corporate revenues, ensuring committed finance to U.S. industry. In the United States, the separation of stock ownership from strategic control during the first decades of this century entailed the transfer of strategic control from owner-entrepreneurs to salaried managers. This transition to organizational control, characterized by organizational integration within the managerial structure and financial commitment on the basis of corporate retentions, provided social foundations for the rise of the United States to international industrial leadership during the first half of this century. During the second half of the century, however, the overextension of the U.S. corporate enterprise into diverse business activities, the challenges of more innovative industrial enterprises abroad, and the transformation of the power of stockholders in the United States eroded the effectiveness of organizational control in U.S. corporations and created growing opportunities for the assertion of market control.

Foundations of Organizational Control

Compared with its own history as well as the current situation in most advanced economies (with the particular exception of Britain), market control presently dominates the governance of big business in the United States. Manifesting market control are the high levels of financial distributions, in forms of both dividends and stock repurchases, of U.S. nonfinancial corporations. Dividends as a proportion of corporate earnings—the payout ratio—increased from about 45 percent in the 1960s and 1970s to over 60 percent in the 1980s and 1990s. Stock repurchases rose even more dramatically. During the 1980s the net equity issues for U.S. corporations became negative in many years. In 1985, stock repurchases were about 22 percent of dividends; in 1989, about 47 percent. From 1990 to 1993 annual stock repurchases averaged about $33 billion, but in 1994 rose to close to $70 billion—33 percent of dividends—and during the first nine months of 1995 were already greater than that amount (*Wall Street Journal* 1995).

Throughout the 1980s, American corporations demonstrated a mounting predilection toward downsizing, with professional, administrative, and technical personnel—so-called white-collar employees—as well as blue-collar workers suffering from the elimination of previously stable and remunerative jobs (Lazonick and O'Sullivan 1996a). To justify such actions, U.S. corporate managers proclaimed that the prime, if not only, responsibility of the corporation was to create value for shareholders. For their success in maximizing shareholder wealth, these strategic managers received ample, often exorbitant, personal rewards, even as most other corporate employees experienced lower earnings and less employment stability. This alignment of the strategic managers of U.S. industrial corporations with public stockholders is now typically regarded as a defining feature of the market-oriented U.S. system of corporate governance.

In historical perspective, however, market control over the allocation of U.S. corporate resources stands out as a recent phenomenon. For most of this century public stockholders exercised little if any control over the allocation of corporate revenues. Ownership of corporate stock was separated from strategic control. Public stockholders accepted this separation because of the liquidity of corporate equities traded on the stock market. This liquidity in turn stemmed from the fact that the market in industrial securities arose as a result of the growth of dominant enterprises in the U.S. economy during the last decades of the nineteenth century. The successful development of these enterprises made possible the rise of a market in industrial securities, not vice versa.

A market in industrial (as distinct from railroad and government) securities in the United States only came into existence in the late 1890s and early 1900s as a number of owner-controlled companies that had grown to dominant positions in their industries in the decades after the Civil War decided to go public (Navin and Sears 1955; Chandler 1977). Launched on the basis of inside capital provided by the entrepreneur, family members, friends, and business associates, a company that transformed itself from a new venture to a going concern in the late nineteenth century did so by reinvesting earnings to build a productive organization. Both the possibility and desirability of a transfer of ownership of the company arose when the managerial organization required to run the enterprise had grown beyond the capacity of a single person or even a small group of partners. The transfer of ownership particularly became a problem when the original owner-entrepreneurs and their backers were ready to retire, as was the case by the 1890s for the post–Civil War generation of entrepreneurs (Navin and Sears 1955, 108). In most cases, the owner-entrepreneur's family members were ill-suited to run a company that relied on organizational learning for its competitive advantage.

The emergence of a market for industrial securities permitted the original owner-entrepreneurs of a highly successful enterprise to sell the company (and often to retire from the industrial scene themselves) while leaving intact the managerial organization that had given the company its competitive advantage as a going concern. Unless the managers themselves were to assume ownership of the enterprise—an ownership transfer that in the most successful companies was generally beyond their collective means—the continued integration of strategy and learning within the enterprise required that stock ownership be separated from strategic control.

In the 1890s and early 1900s initial "public" offerings floated by Wall Street investment bankers went to a relatively small circle of wealthy individuals (including the companies' original owner-entrepreneurs and their families) and financial institutions, particularly insurance companies and the underwriting investment banks themselves. Of the $6.2 billion of industrial common and preferred stock issued during the peak of the Great Merger Movement between 1898 and 1902, 49 percent was privately placed in exchange for the assets or securities of merged companies, and another 45 percent was issued by companies to their own stockholders as dividends or for cash or for unknown purposes. Only 6 percent was sold to the general public (Nelson 1959, 94).

To ensure themselves an income from industrial securities that might be difficult to market, these early portfolio investors favored preferred stocks or corporate bonds rather than common stocks. Indeed, in many initial offerings, common stocks were distributed as a bonus to the purchasers of preferred stocks or to the promoters and investment banks for their services (Flynn 1934, 140). As the market for industrial securities developed, these stockholders were able to sell off some of their preferred and common stocks to the public.

Over time, as the companies listed on the New York Stock Exchange continued to thrive, and as wealthy individuals and institutional investors sold off some of their corporate stocks, stockholding became more dispersed and the threat of outside interference by substantial stockholders decreased (Means 1930). The sale of Liberty bonds during World War I brought the savings of a whole new tier of American households into the securities markets. After the war, Wall Street captured much of these savings through sales of preferred stocks, which were marketed as having the security of bonds (Carosso 1970, 250; U.S. Bureau of the Census 1976, 1005–6).

During the 1920s many of the leading industrial companies made company stock available for purchase by employees, and by 1928 there were more than 800,000 employee stockholders (Means 1930, 568). Companies also sold stock to customers, resulting in the addition of one million new stockholders between 1920 and 1928 (Means 1930). During the 1920s the marketing of stock itself became a highly developed business, with institutional forms of stockholding such as the investment trust becoming popular among small-scale investors (Carosso 1970, chap. 14). Stock splits also became common as a way of making stocks more accessible to households further down the income scale (Dewing 1934, 98). In 1900, there were an estimated 4.4 million stockholders on the books of U.S. corporations, holding an average of 140 $100 par value shares. By 1928 there were 18 million stockholders holding an average of 51 shares (Berle and Means 1932, 56).

The stock market boom of the late 1920s brought new people into the market, many of them borrowing to buy stocks on margin in attempts to get rich quick. In 1927 an unprecedented $1.7 billion of new stock was issued, but that amount doubled the next year and again in 1929. In the process, common stocks gained wide acceptance. In 1927 common stock issued was only 65 percent of preferred stock issued; in 1929, 300 percent (U.S. Bureau of the Census 1976, 1006).

As stockholding became increasingly dispersed, the possibility diminished that any group of stockholders could challenge managerial control of corporate resources. Most corporate stock, whether preferred or common, carried voting rights, but the dispersion of stockholding made it difficult for any small group of stockholders to use these rights to challenge managerial control. When dividend payments were missed preferred stockholders, behaving more like creditors than speculators, tended to scrutinize managerial actions and performance more closely than did common stockholders. As a result, over the first decades of the century corporate managers tended to dilute the power of preferred stockholders by granting common stockholders more votes per dollar of stock. In the 1920s U.S. corporate managers found that they could dilute shareholder power even more explicitly through the issue of nonvoting stock (Sears 1929, 90–91). Subsequently, when the New York Stock Exchange (NYSE) required that listed stock carry voting rights, the result was not to increase shareholder power but rather, by maintaining public confidence in the holding of corporate stock, to foster the further dispersion of stockholding, thus making it all the more difficult for a small group of stockholders to challenge managerial control.

Meanwhile, from the nineteenth century, the evolution of corporation law had strengthened the right of managers to exercise control, not because managers represented a special interest group, but as part of a general process by which the U.S. laws of property supported developmental change, even when such change conflicted with existing property rights (Horwitz 1977, 1992; Sklar 1988). The most important power accorded to corporate managers by statutes and through the courts was the right to declare or withhold dividends (Hurst 1970). In general, the law was not eager to interfere with managerial discretion.

Neither were public stockholders, whose main interest in holding corporate stock was liquidity. Companies listed on the NYSE—which quickly became the exchange of preference for all of the leading U.S. industrial enterprises—had established themselves as dominant enterprises in their particular industries with the capacity to generate high levels of profit on a regular basis. Once public, these high levels of profits made possible continuous dividend payments, which convinced stockholders of the liquidity of their stock. By refusing to cut dividends except under the most dire circumstances, corporate managers ensured that stockholders would not challenge their control over the allocation of corporate revenues.

Wall Street helped to create confidence in the liquidity of corporate stocks by identifying and actively promoting companies that had already acquired the productive base to generate a consistent stream of profits. Ever more stringent requirements for listing on the New York Stock Exchange built public confidence in the stock market, which, in bringing in new buyers of stocks, added further to the liquidity of the market (Hurst 1970).[2] From the second decade of the century, public confidence was further bolstered by the securities ratings services of Moody's and Standard and Poor's, whose own businesses were based on their reputations for impartiality and credibility (Belkaoui 1983, 10–11).

Most important, however, in laying the foundation for a highly liquid market in industrial stocks was the emergence from the last decades of the nineteenth century of a large number of dynamic industrial enterprises that, through the superior development and utilization of productive resources, had gained distinct competitive advantage in an era before a market in industrial securities even existed. These companies—many of which still maintain dominant market shares—had by the 1920s made the United States the most powerful industrial nation in the world.

The stock market did not serve as a source of funds for long-term business investment. When an enterprise went public, the stock market was the instrument for the separation of stock ownership from strategic control over internally generated corporate revenues. As owner-entrepreneurs of dominant going concerns retired, cash and securities in hand, they left investment decision making under the control of the most able, energetic, and visionary of the career managers whom they had recruited to these companies during the previous decades.

The purpose of the ownership transfer was often also to enable one company to acquire another company. Typically, the acquiring company issued new stock in its corporation to be exchanged for the existing stock of the acquired company, whose stock was then retired (Koch 1943). In the aftermath of the acquisition, the acquiring company may have made substantial investments in the acquired company, but the issue of equity did not provide the source of such investment financing.

Funds raised through equity issues were also to be used to restructure corporate balance sheets by using the proceeds from the sale to pay off debt, build cash reserves, or finance acquisitions. Such was the case in the speculative boom of the late 1920s when companies realized that windfalls could be obtained through the sale of equities to the public, who were willing to buy stock at prices that were wildly out of line with the long-run earning power of the underlying corporate assets (Koch 1943, chap. 6).

Throughout the twentieth century, corporate retentions, sometimes leveraged with money raised through long-term bond issues, not equity issues, have been the main sources of funds for business investment. For a sample of 84 large manufacturing corporations, the total amount of funds that they retained over the period 1921 through 1929 just equaled their total fixed capital expenditures (Koch 1943, 81). More recently, for the period 1970 through 1985, retained earnings were 67 percent of the gross sources, and 86 percent of net sources, of funds for U.S. nonfinancial corporations. For the same period, new equity issues represented 0.8 percent of the gross sources and 1.1 percent of the net sources of funds, for U.S. nonfinancial corporations (Hall 1994, 139).

When companies have issued stock for the purpose of financing investment in new productive assets, it has usually been in the form of preferred stock, often with restricted voting rights. In terms of secure prospects of future returns, a preferred stock is much closer to a corporate bond than a common stock. For going concerns that have the option to finance long-term investment on the basis of stocks or bonds, the use of stocks is expensive, less because of high transaction costs involved in equity financing than because an attempt at equity financing signals potential portfolio investors that corporate management does not have the confidence that the company could meet the debt-service requirements of a bond issue. For a company that has access to bond financing to instead choose to issue equity would, therefore, be self-defeating.

Throughout most of this century, U.S. public stockholders had no direct input into the decisions to allocate corporate revenues. Nor did they hire, fire, reward, or punish the corporate managers who made these decisions. Ostensibly the board of directors represented the interests of stockholders in these matters. But it is well known that, historically, the top managers of the U.S. corporation, not the stockholders, have chosen the board of directors, and for stockholders to mount a proxy contest to replace top management was very expensive (Lorsch and MacIver 1989).

In historical perspective, the lack of control of public stockholders over retained earnings of industrial corporations was not imposed on them by corporate managers or government regulators, as some have contended.[3] The market in industrial securities evolved in the United States to effect the separation of stock ownership from strategic control because it offered American households liquidity but did not require commitment. Once the market in industrial securities came into existence, American households were willing to hold stocks in publicly traded corporations only because their "ownership" stakes did not entail any commitments of their time, effort, or additional funds to ensure the success of the company. A general willingness to leave control over the allocation of corporate revenues with managers stemmed in part from the prior revenue-generating successes of the publicly listed corporations under managerial control and in part from the limited liability protection that public stockholders enjoyed. But, more fundamentally, this abdication of control derived from the confidence of public stockholders that the stocks they held were liquid and hence could be sold on the stock market at any time.

In U.S. industrial development, therefore, the emergence of dominant enterprises gave rise to stock markets, not vice versa. The willingness of public investors to hold industrial stocks without the ability to exercise control over the allocation of corporate revenues meant that the ownership rights that adhered to these stocks actually enhanced the organizational control of corporate managers. In the 1920s and 1930s, and in many companies well beyond, with strategic managers integrated into the organizational learning processes of their enterprises, the financial commitment made possible by organizational control provided an essential foundation for the pursuit of innovative investment strategies.

Evolution to Market Control

Yet, the very growth of the enterprise that the separation of ownership from control made possible created conditions that, in the presence of the ideology that public stockholders are the true owners of the enterprise, tended to separate strategic managers from organizational learning. The separation of stock ownership from strategic control, as it occurred in the United States, encouraged both the continuous growth of U.S. industrial corporations and the centralization of strategic control so that these corporations grew larger and larger as unitary strategic entities. As these corporations grew through expansion, extension, and diversification, they often reaped "cumulation advantages" by building on existing capabilities (O'Sullivan 1996, chap. 4). But they also faced the "cumulation disadvantages" of organizational segmentation, which are not inherent in, but more likely to occur with rapid and large-scale enterprise growth. Specifically, the expansion of the enterprise within markets, across vertical activities, and into new markets could lead to the separation of strategic managers from organizational learning processes—that is, strategic segmentation.

In the 1920s U.S. industrial corporations undertook a wave of acquisitions for purposes of both vertical integration and diversification (U.S. Bureau of the Census 1976, 914). Unlike the turn-of-the-century merger movement that contributed to the rise of a market in industrial securities, the acquisition movement of the 1920s could make use of a highly liquid stock market. Stockholders of the acquired firms were willing to accept the stock of the acquiring corporations as payment for their equity holdings.

The growth of U.S. industrial corporations slowed substantially during the 1930s, but the major corporations kept their managerial organizations largely intact and expanded their research and development capabilities (Mowery 1986, 191–92). Armed with these capabilities, corporations such as General Motors, Ford, General Electric, Westinghouse, and Du Pont received huge government subsidies for research, development, and expansion during World War II and then became integral to the postwar military-industrial complex. During the 1950s, mergers and acquisitions became prominent once again. In 1947 the 200 largest U.S. manufacturing corporations held 47 percent of all U.S. corporate assets; in 1957 56 percent, and in 1967 59 percent; (Hay and Untiet 1981, 165). Unless strategic decision making could be effectively decentralized within these corporations, strategic segmentation was likely to result.

In 1962 Alfred D. Chandler documented the emergence and diffusion of the multidivisional structure from the 1920s through the 1950s. By permitting administrative decentralization, this organizational structure was supposed to permit the corporation to diversify into many new businesses without succumbing to strategic segmentation (Chandler 1962). But Chandler's conceptualization of the corporate head office as the realm of strategic decision making and the corporate divisions as the realms of operational control already contemplated the segmentation between strategy and learning that in the 1940s and 1950s had begun to afflict some of the largest, and previously most successful, corporations (O'Sullivan 1996, chap. 5).

The conglomerate movement of the 1960s both reflected and exacerbated that segmentation. These mergers and acquisitions entailed conglomeration of lines of business that had no technological or market relations to one another. In 1948 through 1955 only 10 percent of acquired assets were in the "pure conglomerate" category; in 1964 through 1971 35 percent; and in 1972 through 1979 46 percent. By 1972 through 1979, horizontal or vertical acquisitions in the same line of business had fallen to 23 percent of all assets acquired, down dramatically from 49 percent in 1948 through 1955.

For the top 200 U.S. manufacturing companies ranked by sales, the mean number of lines of business rose from 4.8 in 1950 to 10.9 in 1975. Of the 148 companies of the 200 largest in 1950 that survived until 1975, the mean number of lines of business was 5.2 in 1950 and 9.7 in 1975. Among the best known conglomerates were Beatrice Foods (290 acquisitions between 1950 and 1978), W. R. Grace (186), International Telephone and Telegraph (163), Gulf and Western Industries (155), Textron (115), Litton Industries (99), and LTV (58) (Ravenscraft and Scherer 1987, 30, 32, 38, 39).

Propounding the ideology that a well-trained general manager could manage anything, the conglomerate movement glorified strategic segmentation. In acquiring companies and consolidating financial decision making in the head office, the conglomerate typically stripped control from those who had been the strategic managers of the acquired businesses. The conglomerates often initially retained these former top managers as divisional heads, but failure to meet financial performance targets often led to their replacement by someone from the head office who had scant knowledge of the organizational learning required to compete effectively on product markets (Holland 1989; Lazonick and West 1995).

As a result, the conglomerate movement of the 1960s gave way to large-scale divestiture of businesses in the 1970s. For the first time, the buying and selling of companies became a major financial business (Chandler 1990, 624). As Wall Street turned merger and acquisition activity into an end in itself, companies with low price-earnings ratios became targets for takeover because the acquisition of such companies produced a one-shot increase in earnings per share of the conglomerate. These increases in earnings both profited stockholders and augmented the financial capacity of the conglomerate to make more acquisitions.

By the late 1960s, conglomerate financing had graduated from cash to stock to debt, with the debt-equity ratio in U.S. manufacturing rising from 0.48 in 1965 to 0.72 in 1970 (U.S. Bureau of Census 1976, 928). Then, in the 1970s and early 1980s, came deconglomeration as about one-third of the acquisitions made in the conglomerate movement were sold off, typically under conditions of financial duress (Ravenscraft and Scherer 1987, 190). In 1975 and 1976, divestitures were actually greater than announced mergers and acquisitions (Merrill Lynch Advisory Services 1994, 120).

From the perspective of productive performance, the divestitures that followed conglomeration had the potential for rectifying the problems of strategic segmentation that the conglomerate movement had exacerbated. But the failure of the conglomerate movement also laid the foundations for the rise of a new financial instrument—the high-yield or junk bond—that during the 1970s further increased the incentive and ability for Wall Street to treat productive enterprises like financial assets. Far more than even the debt-financed conglomeration of the late 1960s, the use of junk bonds for buy-outs and takeovers enforced financial liquidity on U.S. industrial corporations, thus subjecting them to market control.

In the early 1970s a portion of the debt incurred in the conglomerate movement, either to acquire other companies or to fend off acquisition, emerged as fallen angels: investment-grade bonds that were now rated below investment grade (Taggart 1988, 9; Bruck 1988, 27, 37–38, 44). In 1973, when the value of these bonds stood at $8.2 billion, Michael Milken, an employee of the Wall Street firm of Drexel Burnham, began to convince institutional investors that they should hold these higher risk securities to get higher yields.

He convinced enough; over the next few years he created a liquid market in junk bonds (Bruck 1988, chap. 7).

In 1977, junk bonds represented only 2.5 percent of corporate bonds outstanding, and in 1982 3.8 percent. But by 1985 9.0 percent of all corporate bonds were junk (Taggart 1988, 9). Underlying the increase in junk bonds from the late 1970s were new issues, a business that Milken's Los Angeles office of Drexel Burnham Lambert quickly turned into its own (Taggart 1988, 8; Bruck 1988; Stewart 1991). Newly issued junk bonds totaled $8 billion for the six years 1977 through 1982, with issues in the last year at $2.7 billion. In 1983 alone, the value of junk bond issues leaped to $8 billion, climbing to a peak of $34.3 billion in 1986 (Perry and Taggart 1990, 187).

The newly issued junk bonds financed divisional managers of a conglomerate enterprise to separate their division from the conglomerate structure. By placing divisional managers in positions of strategic control, these buy-outs created the possibility for the reintegration of strategy and learning. But with intensifying global competition, the disruptions to the organizational learning processes that conglomeration had typically entailed appear to have severely diminished the cumulation advantages inherent in the divisional buy-outs, once they had gained strategic independence (Holland 1984).

There is evidence that the debt-service requirements of these divisional buy-outs resulted in significant cost-cutting (Lichtenberg and Siegel 1990). In trimming the fat of day-to-day operations, many companies found that they also had to cut out the bone of developmental investment. The debt that financed the buy-outs did not fund investment in new productive assets but only transferred claims over the returns to existing assets, with the difference that, by loading the company up with debt, the junk-bond-financed buy-out made it imperative for the enterprise to pay out earnings rather than retain them for new productive investment.

During the 1980s those who ran the Wall Street firms that made money from financing buy-outs were well aware of the advantages of debt service in enforcing liquidity on corporate management, and they were also eager to portray the imposition of financial discipline on corporate managers as the solution to the poor performance of American industry. Frederick Joseph, CEO of Drexel Burnham Lambert during the Milken era, summed up the view from the Street:

> Increased debt has important consequences for management. It reduces discretion in spending free cash flow. Instead of pouring free cash flow into perks or unproductive investments, management is forced to direct cash flow to debt service, effectively returning it to the investing public.

> 'Debt creation without retention of the proceeds of the issue enables managers to effectively bond their promise to pay our (sic) future cash flows,' notes Harvard economist Michael C. Jensen. 'Thus debt can be an effective substitute for dividends.' (Joseph 1990, 123–24)

By 1990, when these words were published, the use of junk bonds had long since been transformed from financing divisional buy-outs to financing hostile takeovers of entire corporations. Corporate raiders of the 1980s were often handpicked by Michael Milken to launch a takeover; they then relied on his network of institutional investors (including savings and loan companies) to commit to buy the junk bonds that the acquired company would issue when it was taken over as the means to buy up the stock that allowed it be taken over! In 1986 there were 76 public company buy-outs—20 percent of all public takeovers—at a real average value (in 1988 dollars) of $303.3 million. The average real value of 47 company buy-outs in 1987 and 125 in 1988 was around $480 million, about three times the average value of the divisional buy-outs in those years (Jensen 1989, 65; Merrill Lynch Advisory Services 1994, 95).

The purpose of a high-value public-company buy-out, such as the much-publicized KKR takeover of RJR Nabisco, was, to use Michael Jensen's apt phrase, "to disgorge the free cash flow" from companies that had allegedly "matured" (1986, 1989).[4] The attempt to transform committed finance into liquid finance sent stock prices up when the target was put in play, as did the speculative bidding that occurred as the possibility of takeover loomed near. To pay for the high cost of the takeover, and the high rates of interest that were paid for junk-bond financing, divisions were sold off, and the cash flow of the company was made as free as possible from other claims—such as those of long-time employees, suppliers, and customers (Shleifer and Summers 1988).

What financial interests bent on financial liquidity count as free cash flow may be resources that, with appropriate organizational integration, could be reinvested in organizational learning. During the 1980s control over corporate revenues shifted dramatically in favor of such financial interests. In addition to debt-enforced liquidity, the decade saw a substantial increase in the ability of institutional investors to get high yields on corporate stock.

Households held 90 percent of all corporate equity in 1952, but only 48 percent in 1994, and more recently undoubtedly even considerably less, as, during the first half of 1996, record sums—$50 billion dollars in January and February alone—flowed into mutual funds to reap the returns of the stock market boom (Wyatt 1996). Collectivizing the savings of households, from 1952 to 1994 the share of outstanding corporate stocks held by pension funds increased by 25 percentage points and by mutual funds by 10 percentage points. In the 1980s and 1990s marked increases in dividend:profit payout ratios combined with unprecedented corporate stock repurchases and corporate downsizing have supported the high yields on corporate stock. In addition, with inflation defeated—largely because of the much diminished power of the American labor movement as well as the pressures of foreign competition—real bond yields became markedly higher in the 1980s and 1990s than they had been in previous decades (U.S. Congress 1996, 343, 360). Also pushing up bond yields were the deregulation of financial markets from the late 1970s as well as the integration of the bond and stock markets through the investment portfolio strategies of the institutional investors.

Through the collective power of institutional investing, American households—at least the 45 percent who have pension coverage (Ghilarducci 1992, 3)—can now put pressure on companies to get their stock yields and stock prices up. During the 1960s the mutual funds, with about 85 percent of their assets in stocks, increased their control over outstanding stock to over 4 percent and played an important arbitrage role in the conglomeration movement by buying up large blocks of stock that were rumored to be in play and selling them to corporate raiders at higher prices (Editors of *Fortune* 1970, 142). In 1975 the institutional investors, now faced by inflation and low securities yields, pressured Wall Street to end fixed commissions on trading, setting the stage for a major increase in the volume of trading through the churning of investment portfolios, and intensifying the shift of Wall Street from investment banking to securities trading (Vietor 1987, 42–45; Auletta 1986; Carrington 1985). The participation of institutional investors made it possible for Milken to create the junk-bond market in the 1970s and to use it to launch hostile takeovers in the 1980s.

By the mid-1980s institutional investors could have a direct effect on corporations. The top managers of American companies have been increasingly open—and amply rewarded for acceding—to institutional investors' demands. In the aftermath of the October 1987 stock market crash, major institutional investors, led by California State Public Employees Retirement System, began to engage in relational investment to get companies to take actions that would increase the value of their stockholdings.[5] As a result, the S&P index only declined by 7 percent in 1988, bounced back by well over 21 percent in 1989, and has scarcely looked back since. In replacing organizational control by market control, the revolutions in production and finance that were taking place in the 1970s have in the 1980s and 1990s drastically undermined the social conditions for innovation and industrial development in the United States.

Corporate Governance and International Competition

The finance of industrial development cannot be understood in the absence of an explanation of the sources of industrial development. Recently, proponents of market control have pointed, with justification, to managerial abuse and ill-informed allocations of corporate resources as the rationale for "creating value for shareholders" through dividend payments, stock repurchases, and work-force downsizings (Jensen 1989). Problems of corporate governance are real. Especially in the United States, the top corporate executives are grossly overpaid relative to other employees in their organizations and to their top management counterparts in other advanced economies (Crystal 1991). And over the past few decades, under a system in which strategic decision making has increasingly been concentrated at the top of large enterprises, American corporations have, in many industries in which they used to be world leaders, lost substantial market shares.

The proponents of market control define the source of the problem as agency costs that, with stock ownership separated from managerial control, take the form of "managerial discretion." They then assume that managerial discretion can only take the forms of managerial perquisites and unproductive investments—resources that managers allocate to advance their own selfish interests. The solution is the "market for corporate control" that pressures incumbent managers to "create value for shareholders" or else lose their positions of allocative control.

What the proponents of market control see as a solution to the dissipation of resources by management we see as part of the problem. Strategic managers need to have discretion if investments in developing and utilizing productive resources are to be made that result in sustained competitive advantage for their enterprises and sustainable prosperity for the economy. But who these decision makers are, how they make their decisions, and whom they seek to benefit have profound impacts on whether these companies invest for the future or live off the past.

The historical evidence shows that the problem for innovation and industrial development is not that strategic managers have discretion, but that they have become too separated from the organizations in which the development and utilization of productive resources occurs. The market for corporate control tends to exacerbate this strategic segmentation. In the presence of a powerful market for corporate control, the use of stock-based rewards aligns the interests of strategic managers with public stockholders, making it all the more certain that the integration of strategy and learning will not occur (Lazonick 1992).

Once one recognizes the importance of organizational learning to the development and utilization of productive resources, one cannot avoid the fact that the most important investments that an innovative enterprise makes are in human resources, not physical resources. Although in common parlance business executives will say that their human assets are their companies' most valuable assets, in corporate law and in accounting practice human capabilities are not treated as corporate assets because people cannot be owned. Investments in human resources are counted as *expenses,* as are the returns to human resources that take the forms of higher incomes, better benefits, and more stable employment. The conventional concept of property on which these accounting practices are based, however, ignores the collective assets and collective returns that are the essential realities of innovative enterprise (O'Sullivan 1996, chap. 7).

The problem of corporate governance and industrial development is not resolved by simply advocating, as has recently been done in both the United States and Britain, that industrial corporations be run for other stakeholders—especially employees—besides stockholders (Reich 1996a). The danger is that different groups who can lay claim to shares of corporate revenues will, as has increasingly been the case with stockholders, extract corporate revenues whether or not their contributions to the generation of these revenues makes these returns possible on a sustainable basis. The result of the creation of a stakeholder society might be to increase the propensity for major industrial enterprises and the economy in which they operate to live off the past rather than invest for the future.

The historical evidence shows the importance of collective organization, and the integration of strategic decision makers into this organization, for successful industrial development. In the United States during the first half of the century this collective organization took the form of the managerial revolution—a transformation in business organization that provides a major explanation for the industrial success of the United States during this century. Even so, by the late twentieth century, organizational integration in the United States has itself fallen short of internationally competitive standards. As we argue elsewhere in comparisons of the U.S. experience with industrial development in Japan and Germany, the weakness of American industry in international competition is that organizational integration has not been extended to broader groups of people both within and across industrial enterprises to meet the evolving requirements of industrial innovation (Lazonick and O'Sullivan 1996a, 1996b, 1997b).

The ideology that the "shareholder" is the "principal" for whom the industrial corporation is run helps to ensure that such organizational transformations will not take place. This ideology places a premium on economic performance that reaps the benefits of prior investments in productive capabilities while ignoring the new investments in organizational learning that can potentially generate greater returns for more people in the future.

NOTES

1. For our comparison of the United States with Britain, see Lazonick and O'Sullivan 1997a.
2. See also R. Michie 1987.
3. For example, Roe 1994.
4. See also Burrough and Helyar 1990; Anders 1992; Johnston 1986; Bruck 1988.
5. See Blair 1995, chap. 5.

REFERENCES

Anders, G. 1992. *Merchants of Debt: KKR and the Mortgaging of American Business.* New York: Harper Collins, Basic Books.

Auerbach, A. J., ed. 1988. *Mergers and Acquisitions.* National Bureau of Economic Research project report. Chicago: University of Chicago Press.

Auletta, K. 1986. *Greed and Glory on Wall Street.* New York: Warner Books.

Belkaoui, A. 1983. *Industrial Bonds and the Rating Process.* Westport, Conn: Greenwood.

Berle, A. A., and G. C. Means. 1932. *The Modern Corporation and Private Property.* New York: Macmillan.

Blair, Margaret M. 1995. *Ownership and Control: Rethinking Corporate Governance for the Twenty-First Century.* Washington, D.C.: Brookings Institution.

Bruck, C. 1988. *The Predators' Ball: The Junk-Bond Raiders and the Man Who Staked Them.* New York: BDD Promotional Book Company.

Burrough, B. and J. Helyar. 1990. *Barbarians at the Gate.* New York: Harper Collins.

Carosso, V. P. 1970. *Investment Banking in America.* Cambridge, Mass: Harvard University Press.

Carrington, T. 1985. *The Year They Sold Wall Street.* Boston: Houghton Mifflin.

Chandler, A. D. Jr. 1962. *Strategy and Structure: Chapters in the History of the Industrial Enterprise.* Cambridge, Mass. MIT Press.

———. 1977. *The Visible Hand: The Managerial Revolution in American Business.* Cambridge, Mass: Harvard University Press.

———. 1990. *Scale and Scope: The Dynamics of Industrial Capitalism.* Cambridge, Mass: Harvard University Press.

Crystal, G. 1991. *In Search of Excess: The Overcompensation of American Executives.* New York: W. W. Norton.

Dewing, A. S. 1934. *A Study of Corporation Securities, and Their Nature and Uses in Finance.* New York: The Ronald Press Co.

Editors of *Fortune.* 1970. *The Conglomerate Commotion.* New York: Viking Press.
Flynn, J. T. 1934. *Security Speculation: Its Economic Effects.* New York: Harcourt Brace.
Ghilarducci, T. 1992. *Labor's Capital: The Economics and Politics of Private Pensions.* Cambridge, Mass: MIT Press.
Hall, B. H. 1994. Corporate Restructuring and Investment Horizons in the United States, 1976–1987. *Business History Review* 68, no. 1: 110–43.
Hay, G., and C. Untiet. 1981. Statistical Measurement of the Conglomerate Problem. In *The Conglomerate Corporation: An Antitrust Law and Economics Symposium,* ed. R. D. Blair and R. F. Lanzillotti. Weston: Oelgeschlager, Gunn, and Hain.
Holland, M. 1989. *When the Machine Stopped: A Cautionary Tale from Industrial America.* Boston: Harvard Business School Press.
Horwitz, M. 1977. *The Transformation of American Law, 1780–1860.* Cambridge, Mass: Harvard University Press.
———. 1992. *The Transformation of American Law: The Crisis of Legal Orthodoxy, 1870–1960.* New York: Oxford University Press.
Hurst, W. J. 1970. *The Legitimacy of the Business Corporation in the Law of the United States, 1780–1970.* Charlottesville, Va: University Press of Virginia.
Jensen, M. C. 1986. The Agency Cost of Free Cash Flow, Corporate Finance, and Takeovers. *American Economic Review* 76, no. 2: 323–29.
———. 1989. The Eclipse of the Public Corporation. *Harvard Business Review* 67, no. 5: 61–74.
Johnston, M. 1986. *Takeover: The New Wall Street Warriors.* New York: William Morrow.
Joseph, F. H. 1990. A Wall Street View of the High-Yield Debt Market and Corporate Leverage. In *The High-Yield Debt Market: Investment, Performance, and Economic Impact,* ed. E. I. Altman. New York: McGraw-Hill.
Koch, A. R. 1943. *The Financing of Large Corporations, 1920–39.* Cambridge: National Bureau of Economic Research.
Lazonick, W. 1992. Controlling the Market for Corporate Control. In *Entrepreneurship, Technological Innovation, and Economic Growth: Studies in the Schumpeterian Tradition,* eds. Frederic M. Scherer and Mark Perlman. Ann Arbor: University of Michigan Press.
Lazonick, W., and J. West. 1995. "Organizational Integration and Competitive Advantage: Explaining Strategy and Performance in American Industry. *Industrial and Corporate Change* 4, no. 1: 229–70.
Lazonick, W., and M. O'Sullivan. 1996a. Corporate Employment and Corporate Governance: Is Prosperity Sustainable in the United States? Report to the Jerome Levy Economics Institute, October.
———. 1996b. Organization, Finance, and International Competition. *Industrial and Corporate Change* 5, no. 1: 1–49.
———.1997a. Finance and Industrial Development. Part I: The United States and the United Kingdom. *Financial History Review* 4, no. 1: 7–29.
———.1997b. Finance and Industrial Development: Evolution to Market Control. Part II: Japan and Germany. *Financial History Review* 4, no. 2: 117–38.
Lichtenberg, F., and D. Siegel. 1990. The Effects of Leveraged Buyouts on Productivity and Related Aspects of Firm Behavior. *Journal of Financial Economics* 27, no. 1: 165–94.
Lorsch, J., with E. MacIver. 1989. *Pawns or Potentates: The Reality of America's Corporate Boards.* Boston: Harvard Business School Press.

Means, G. C. 1930. The Diffusion of Stock Ownership in the United States. *Quarterly Journal of Economics* 44, no. 4: 561–600.
Merrill Lynch Advisory Services. 1994. *Mergerstat Review.* New York.
Michie, R. 1987. *The London and New York Stock Exchanges, 1850–1914.* London.
Mowery, D. 1986. Industrial Research, 1900–1950. In *The Decline of the British Economy,* ed. W. Lazonick and B. Elbaum. New York: Oxford University Press.
Navin, T., and M. Sears. 1955. "Rise of a Market in Industrial Securities, 1887–1902. *Business History Review* 29, no. 2: 105–38.
Nelson, R. L. 1959. *Merger Movements in American Industry, 1895–1956.* Princeton: Princeton University Press.
O'Sullivan, M. 1996. Innovation, Industrial Development, and Corporate Governance. Ph.D. dissertation. Harvard University.
Perry, K. J., and R. A. Taggart Jr. 1990. Development of the Junk Bond Market and Its Role in Portfolio Management and Corporate Finance. In *The High-Yield Debt Market: Investment, Performance, and Economic Impact,* ed. E. I. Altman. New York: McGraw-Hill.
Ravenscraft, D. and F. M. Scherer. 1987. *Mergers, Sell-Offs, and Economic Efficiency.* Washington, D.C.: Brookings Institution.
Reich, R. B. 1996a. How to Avoid These Layoffs? *New York Times,* January 4: A13.
———. 1996b. Shareholder Values. *Economist* (February 10): 15
Roe, M. 1994. *Strong Managers, Weak Owners.* Princeton: Princeton University Press.
Sears, J. H. 1929. *The New Place of the Stockholder.* New York: Harper and Brothers.
Shleifer, A., and L. H. Summers. 1988. Breach of Trust in Hostile Takeovers. In *Corporate Takeovers: Causes and Consequences,* ed. A. J. Auerbach. Chicago: University of Chicago Press.
Sklar, M. J. 1988. *The Corporate Reconstruction of America, 1890–1916: The Market, the Law and Politics.* New York: Cambridge University Press.
Stewart, J. B. 1991. *Den of Thieves.* New York: Simon and Schuster.
Taggart, R. A., Jr. 1988. The Growth of the "Junk" Bond Market and Its Role in Financing Takeovers. In *Mergers and Acquisitions,* ed. A. J. Auerbach. National Bureau of Economic Research Report. Chicago: University of Chicago Press.
U.S. Bureau of the Census. 1976. *Historical Statistics of the United States.* Washington, D.C.: Government Printing Office.
U.S. Congress. 1996. *Economic Report of the President, 1996.* Washington, D.C.: Government Printing Office.
Vietor, R. H. K. 1987. Regulation-Defined Financial Markets: Fragmentation and Integration in Financial Services. In *Wall Street and Regulation,* ed. S. L. Hayes, III. Boston: Harvard Business School Press.
Wall Street Journal. 1995. Firms Ponder How Best to Use Their Cash. October 16: A1, A9.
Wyatt, E. 1996. Money Keeps Pouring into Market Funds. *New York Times,* March 7.

PART 3

Competitiveness and Technology Policy

CHAPTER 6

Can Technology Policy Serve as Industrial Policy?

Ann Markusen

In the 1990s, U.S. policymakers have formally placed technology policy on the national economic agenda. With accession of Clinton to the presidency, initiatives begun by the Congress, the Department of Defense's Defense Advanced Research Projects Agency, and Bush's science adviser have grown into multibillion dollar programs for dual use and other technologies to be underwritten by the government. These technology policies are the core of what is not yet quite officially acknowledged as an industrial policy. By subsidizing the development costs of a series of "critical technologies," they are anticipated to spawn new products and industries that will make the United States competitive in the future.

In this paper, I review the strengths and weaknesses of technology policies as currently proposed for the United States, relying on both the historical record and some normative and political economy propositions about their formulation. The paper makes three rather speculative arguments. First, the implementation of technology policies without complementary policies aimed at market creation and favorable finance and trade environments is unlikely to succeed. The dearth of evaluative work on past stand-alone technology initiatives and the mixed results of several high-profile projects both in the United States and abroad do not support the optimism of advocates for new investments in critical technologies.

Second, while the subsidization of new technologies is defended as a means of long-term creation of good jobs in the American economy, the institutional arrangements under which subsidization is being pursued do not ensure that gains will translate into domestic job generation. Nor are performance requirements imposed that ensure a taxpayer return on either an employment or revenue basis. At the same time, technology policies are vulnerable to manipulation and capture by a narrow segment of high-tech industry, a possibility that underscores the importance of institutional arrangements that are transparent and democratic.

Third, it is possible that the subsidization of new technologies may result in disruptive and atrophic growth dynamics, by which accelerated rates of innovation destabilize business strategies and aggravate the displacement of workers and communities without creating absorptive capacity elsewhere in the economy. If the manner in which technology enhancement is pursued worsens the income distribution, this potential is increased. A world in which advanced industrialized nations concentrate their energies and resources on winning contests to dominate risky new technologies could also be one in which general levels of income and welfare could simultaneously decline.

Conceptual and Empirical Work on Technology and Innovation

Contemporary enthusiasm for technology policy is based on interpretations of patterns and processes of innovation in industrialized countries over the past 200 years. In general, and at risk of tremendously oversimplifying the story, the record tells us a number of things, some at a macro level, some at the micro level. This record has been codified into a number of propositions about the relationship between R&D, technology, and growth.

As a theory of both evolution of industry and crisis resolution, the Schumpeterian view of creative destruction serves as the basis for the macroeconomic dimension. Schumpeter's entrepreneurial striving and promotion of new innovations serve as the means for the creation of fundamentally new industries and as a way out of troughs in long waves. When markets are saturated and ordinary business practices (advertising, minor product design changes, reorganization of production) fail to reverse flagging sales, Schumpeter's entrepreneurs turn to fundamentally new techniques and products in desperation. Their doing so in times of deep recession accounts for the bunching up of innovations in the troughs of long waves (Mensch 1979).

We can graft onto this theory the product cycle/profit cycle depiction of the methodical rise and fall of industries in the longer run, with distinctive behaviors with respect to innovation and business practices characterizing each stage (Vernon 1966; Markusen 1985; Utterback 1987). The result is a simple version of generations of industries in which a round of fundamental innovation, set off in the trough of a long wave, creates a new ensemble of industries that then carry the economy along through the next wave of growth and development. It is the belief in the inevitability of the periodic decline of one set of industries and its replacement by another that compels contemporary interest in less proven but promising technologies. (It is also the fear, in an analogue with the arms race, that some other nation will develop that technology or industry first, achieving an interim monopoly and developing at the expense of other nations [Markusen 1991]).

At the micro level, the received wisdom on technology and innovation is somewhat more concrete and empirically well documented, if quite complex. First, invention, whether individual or organizational, has been determined to be an activity distinct from innovation. Many quite useful, even revolutionary, inventions were discovered years before they were embodied in a product or service, often remaining on the shelf and failing to benefit their creator. A second finding is that since the second half of the nineteenth century, innovation has become increasingly science-based—reliant upon organized scientific inquiry and a growing body of scientific knowledge rather than emanating from the benches of backyard tinkerers. Third, large corporations have come, in the twentieth century, to act as major innovators, accounting for growing shares of new products, which are designed and developed in corporate laboratories devoted to this purpose. Fourth, because of problems of appropriability, private sector firms may be underinvesting in research and development from a social welfare point of view.[1]

The economics literature on science and technology tells us much about the changing character and institutional context of innovation, but it does not tell us much about the structural features of markets and the larger economic environment which might nurture or retard technological advance and the emergence of new industries. In particular, the interaction between market pull—a pressing need for an innovation that promises a ready demand—and supply-side factors (personnel, capital, know-how) is complex and not easy to unravel. Yet the respective roles of supply and demand factors are of great import in assessing the emergence and success of both private and state-managed innovation. Historically, it is clear that the search for a way to stop the flooding of mines, for a cure for epidemics, or for better materials to make bridges and buildings more reliable played a significant role in the innovation of steam engines, vaccines, and better steels and concrete (Mowery and Rosenberg 1989). Of course, access to a body of scientific knowledge, universities to train scientists and engineers, and sufficient organizational size and resources to carry out the necessary R&D and innovation is key to successful innovation as well. The tricky issue in technology policy is how to insert state aid on both the supply and demand sides in the most effective (and cost-efficient) way to create new rounds of innovation and industry-building.

The Underappreciated Role of Demand

It is to the supply side that contemporary U.S. technology policies are almost exclusively addressed, invoking the neoclassical hypothesis of private underinvestment and less analytically based fears of loss of technological leadership. The success of the American aerospace, communications, and electronics industries—the ACE complex as we dub it elsewhere (Markusen and Yudken 1992)—is attributed to high levels of government support for both basic and applied R&D, as the public sector came to assume a much larger role in research funding following World War II. The specter of a MITI funding long-term R&D initiatives in Japan is invoked as a further imperative for a new U.S. approach, one that rejects the laissez-faire stance that has prevailed outside of the agriculture, defense, and health sectors.

However, the emergence and success of certain sectors in international competition is considerably more complex than is suggested by the implicit cause-and-effect relationship, posited in arguments for a U.S. technology policy, linking state-supported R&D efforts with the superior performance of certain sectors. In fact, the industries that comprise the bulk of American comparative advantage internationally have been the beneficiaries of a fuller set of policies, varying in each case, but generally encompassing trade, finance, and market-creation efforts. Market creation, through either regulation or procurement, especially in youthful industry stages, has been a powerful tool of industrial formation, one whose effects are difficult to separate from those of R&D funding.

This point can be illustrated by reflecting on the emergence of the American aircraft, computer, electronics, communications, instruments, and arms sectors as major net exporters over the postwar period. Each of these industries has enjoyed considerable patronage from the Department of Defense over many decades. In the case of the aircraft industry, this connection began in World War I when the government ordered 14,000 aircraft from a tiny, craft-oriented industry, simultaneously setting up the National Advisory Committee for Aeronautics to further the science and technology of aeronautics and advise the military on purchases. The government also initiated a trade association, the Manufacturers Aircraft Association, in which rivals pooled patents and shared plane-making methods. After the war, the Association along with its clients in the Army Air Corps lobbied successfully for continued government support for the industry through public operation of the air mail system and public funding of airports, flight control, and weather services. From the start, this industry was thoroughly dependent upon government markets and regulation, as well as government-sanctioned cooperation that in other industries would have been in violation of antitrust laws (Markusen and Yudken 1992, 39–42).[2]

The electronics and communications equipment industries are quite similar in their origins. Beginning in the 1930s, with the development of radar and its need for components to modulate the high-speed, high-frequency electrical impulses generated, substantial government moneys went into both the funding of research and the purchase of new products in everything from radar to transistors to semiconductor materials. Up through the early 1960s, the government funded about 85 percent of U.S. electronics research and development and comprised at least 70 percent of the market for semiconductors and 100 percent for integrated circuits. The early decades of computer research and manufacture were similarly dominated by government purchases as well as research funding (Markusen and Yudken 1992, 47–48).[3]

Even in the 1980s, these industries continued to receive substantial patronage from the Department of Defense, although some, like computers and semiconductors, had weaned themselves from heavy reliance on procurement contracts. In 1985, when the Department of Commerce last computed input-output demands of the Pentagon on industrial sectors, the aircraft industry still depended upon the Pentagon for 66 percent of its market (table 6.1).[4] The communications equipment industry, with larger military sales in absolute terms than the aircraft industry and employing over 350,000 workers, relied on the same source for 50 percent of its sales. Although the computer and semiconductor industries still delivered $2.3 and 1.6 billion (in 1977 dollars) in output to the Department of Defense, their dependency had fallen by 1985 to remarkably modest levels.

These industries still rely heavily on the Department of Defense for research funding (Table 6.2). In 1989, the aerospace industry received 82 percent of all funds it spent on R&D from the federal government. It alone accounted for 54 percent of all federal funds spent on private sector manufacturing R&D. The electronics and communications industries received 43 percent of their R&D funds from the government and accounted for another 27 percent of all federal funding. Scientific instruments and computing also enjoyed federal support far in excess of that awarded to the more traditional industrial sectors like steel, chemicals, paper, and textiles.

This powerful combination of R&D funding with procurement gave industries like aircraft, semiconductors, computers, communications equipment, and scientific instruments a great advantage both domestically and internationally. As a group, those with such government patronage increased their shares of manufacturing output in the economy over the postwar period and particularly after, in the late 1970s, as increased world market integration and heightened competitive pressures set in—computers, electronics, instruments, and aircraft topped the list of industry growth rates in GNP share (table 6.3). These industries also flourished in international trade, accounting for the bulk of net positive exports outside of agriculture and chemicals (table 6.4).

One interpretation of this record is that those industries with this cultivated comparative advantage grew at the expense of those industries less

TABLE 6.1. Growth in Military Sales and Dependency, 1980–83 (1977 = 100)

Industry	Military Output, $ billion 1985	Military Share of Output (%) 1980	Military Share of Output (%) 1985
Shipbuilding	5.8	61	93
Ordnance	0.8	79	86
Missiles	5.3	69	84
Tanks	1.1	68	69
Aircraft	11.7	37	66
Communications equipment	15.7	42	50
Machine tools	0.4	8	34
Engineering instruments	0.7	23	28
Optical instruments	0.9	13	24
Electronic components	3.0	16	20
Steel	3.4	6	12
Airlines	3.0	n/a	10
Oil refining	5.2	4	6
Computers	2.3	5	5
Industrial chemicals	2.0	4	5
Semiconductors	1.6	9	5
Automobiles	6.3	4	3

Source: Adapted from Markusen and Yudken, compiled from Henry and Oliver 1987 and unpublished data from the U.S. Department of Commerce, 1987.

TABLE 6.2. Federal vs. Industry Funding for Selected Industries, 1989 (millions of dollars)

Industry	Federal Funds	Industry Funds	Total	Federal as Percentage of Total	Percentage of all Federal Funds
Aerospace	15,647	3,511	19,157	81.7	53.5
Electronics, communications	7,928	10,618	18,546	42.8	27.1
Rubber products	313	930	1,243	25.2	1.1
Autos, trucks, RR (+Tanks)	1,982	9,431	11,413	17.4	6.8
Scientific instruments	991	5,531	6,522	15.2	3.4
Machinery (inc. computers)	1,669	10,457	12,126	13.8	5.7
Fabricated metals	73	732	805	9.1	0.2
Iron and steel	21	601	622	3.4	0.1
Chemicals	381	11,134	11,515	3.3	1.3
Petroleum products	21	2,068	2,089	1.0	0.1
Food and beverage	0	1,172	1,172	0.0	0.0
Paper/pulp	0	1,009	1,009	0.0	0.0
Textiles	0	176	176	0.0	0.0
Manufacturing	29,223	59,648	88,871	32.89	100.0

Source: Aerospace Industries Association of America, Aerospace Facts and Figures, 1989/1990 (Washington, D.C.: AIA, 1989), 104; derived from Battelle data.

TABLE 6.3. Manufacturing Share of Gross National Product, 1979–86

Year	1986	Change in Share 1979–86
Manufacturing	21.87	0.2
Computers, non-electrical machinery	4.06	54.7
Electric and electronic equipment	2.29	21.4
Instruments and related products	0.80	13.5
Aircraft, other non-auto transport	1.31	12.1
Food and kindred products	1.69	–2.0
Printing and publishing	1.14	–3.3
Chemicals and allied products	1.60	–7.8
Lumber and wood products	0.58	–11.6
Apparel and other textile products	0.53	–14.5
Motor vehicles and equipment	1.20	–20.1
Petroleum and coal products	0.71	–21.5
Primary metal industries	0.94	–39.6

Source: Adapted from U.S. Department of Commerce, Bureau of Economic Analysis, National Income and Product Accounts (1987), tables 6.1, 6.2; U.S. Congress 1988, table 3.

TABLE 6.4. Trade Balance in Selected Manufacturing Industries, 1987 (billions of dollars)

Industry	1987
Total manufacturing	–$137.7
Aircraft and other transportation equipment	12.5
Chemicals	9.6
Professional, scientific, and control instruments	3.0
Military arms, ammunition, vehicles	2.0
Computing and office machinery	1.0
Industrial machinery	–6.7
Semiconductors and other electrical equipment	–7.0
Iron and steel	8.5
Telecommunication and sound reproducing equipment	–15.6
Wearing apparel and accessories	–16.9
Motor vehicles	–53.3

Source: Adapted from data from the U.S. Department of Commerce, Office of Trade and Investment Analysis; U.S. Congress 1988; table 12.

favored as a result of a depletionary effect whereby talent and capital are channeled away from civilian industry toward the beneficiaries of this "quiet industrial policy" (Markusen and Yudken 1992, 58–64).[5]

The fusion of R&D funding with government procurement in the history of these new industries makes it difficult to argue for the efficacy of the one without the other. Government procurement without incentives to innovate or adapt technology does not necessarily create new industries—the shipbuilding and tank industries demonstrate that. But R&D funding or technology transfer without some attention to market creation, especially in the infant industry period, may not work either. Several examples come to mind, although no definitive conclusion can be drawn. The Japanese effort to build a commercial aircraft in the 1970s failed in part because of the absence of a broad enough domestic market to make its production economically feasible (Samuels and Whipple 1989; Mowery and Rosenberg 1989). The American attempt to build a supersonic passenger jet (the SST) also failed, its success predicated on a civilian market that did not materialize at the price.

Very little empirical work has been done to sort out the effect of R&D versus government procurement on industry performance. One exception is Lichtenberg's (1987) critique of the view, based on regression analyses, that government-subsidized R&D leverages private sector spending on innovation. However, recent research by Lichtenberg respecifying the model and further empirical work shows that much of the private sector R&D inferred to be matching public sector R&D can be explained by companies undertaking research in anticipation of government sales. When government sales are added into the regression analysis, the explanatory power of government R&D shrinks by about two-thirds.

Market creation can be furthered by one of several mechanisms—procurement, protection, or regulation. Procurement, as we have seen in the U.S. case, has been a major shaper of the aircraft, communications, and electronics sectors. Protected domestic markets have been a major feature of many countries' successful efforts to enter new industries, for example, steel in the United States in the nineteenth century, or autos and electronics in Japan in the postwar period. Regulation is a relatively less costly way of forming markets. In new initiatives in nonfuel vehicles, regulation has been used as a way of ensuring market formation. In California, stiff new air quality controls mandate that 2 percent of all vehicles sold in the state by 1996 be emission-free, 5 percent by the end of the decade. This, combined with federal funds for battery research and prototype development, has encouraged big and small firms, in both auto and aerospace industries, to explore electric and other non-fuel car production.

The conclusion to be drawn from this discussion is that government shaping of markets may be as important as supply-side incentives (if not more) in the initial seeding of new industries. (Trade and finance policies are important, also, and worthy of their own separate treatments.) Market shaping, because it is expensive and carries opportunity costs, whether imposed on taxpayers or consumers, is best done in the early stages of a new technology or product. Research on the changing nature of spin-offs from military-industrial research suggests that government sponsorship during early stages of generic technological advance, as in nuclear energy, electronics, and computers, promises greater commercial spinoffs, whereas where government missions and R&D funding continue to dominate in later developmental stages, civilian utility will be undercut by increasing orientation to esoteric military technologies and a corporate preference for serving relatively less-risky and better-known government markets (Stowsky 1986; Reppy 1989). Furthermore, a justification for market shaping, which is based on either market failure or public interest grounds, is superior to the pure invocation of "competitiveness." What made defense such a powerful underwriter of innovation and a reliable market for incipient high-tech industries was the consensus that national defense was in the public interest, providing a public good not achievable through purely market mechanisms. I return to this point below.

Normative Issues in Science and Technology Policy

In the recent debate, proponents argue that new government-financed initiatives in technology are in the public interest and assert that quite dramatic contributions to the economy will ensue. The following statement, for instance, is made in an OTA report on technology and conversion from military to civilian production.

> Critical industries and technologies, by definition, make disproportionately large contributions to national well-being through such things as creation of larger than average numbers of highly skilled, well-paid jobs; the promise of productivity or product improvement in many industries; and in many cases, fast-growing markets here and abroad. Yet, many fear that, without new initiatives to advance critical industries and technologies, market signals and current government programs alone are insufficient to assure that American companies maintain prominent places among the world's best competitors. (1993)

The statement makes many claims that on reflection are dubious or are, at most, hypotheses without definitive empirical validation.

To begin with, the definition of critical industries and technologies as those that create large numbers of good jobs begs the question as to how these industries and technologies will be identified. Job creation is a future, rather than a present or past, product of such initiatives and cannot a priori be relied upon to choose targets. In both private and public sectors, new initiatives have often ended up as spectacular and costly failures. Although such failures may be acceptable in return for achieving substantial breakthroughs on some other front, advocates must nevertheless have a proposal for how critical industries/technologies shall be chosen, as well as establish some decision criteria involving cost/benefit analysis on whether to pursue or liquidate projects that are not panning out. As the authors of *Beyond Spinoff* put it:

> Today's ubiquitous lists of "critical technologies" say little about what the government should do, or indeed whether more government support is necessary at all. (Alic, et al. 1992, 371)

Remarkably little attention has been paid by researchers or technology policy experts to the failures and what might be learned from them. Again, in a remarkable statement from the same OTA (1993) report: "It might be wise to examine the conduct and structure of past technology initiatives, particularly successful ones, for some guidance in the preparation of new ones." Research on the failures is implicitly suggested to be less promising than that on the success stories. Yet without such insight, how can we fashion an effective program of government intervention?

Failures in publicly supported programs have been numerous and costly. Examples abound in military-based efforts—a recent example is the Brilliant Pebbles component of Star Wars. Civilian programs have had their shares of failure, too. One is the SST, noted above—the European investment in the Concorde has similarly not paid off commercially. The American manned space program has been heavily criticized as a great waste of money, critics arguing that unmanned space programs could have easily met security and scientific needs much more economically. DOE's heavy investment in the commercialization of nuclear power, with its safety and waste disposal problems, has been judged in retrospect as a mixed bag (Nimroody, Hartung, and Grenier 1988). Even the much-admired Japanese have had their failures. MITI's efforts to develop a commercial airplane in the 1970s, for instance, was a spectacular and expensive failure (Samuels and Whipple 1989).

Even where policies succeed in creating competitive new industries, the costs of pursuing them often exceed the benefits. In a study of commercial aircraft innovation, Mowery (1986) estimates a negative social rate of return on the substantial public and private R&D investments in aircraft performance through the early 1980s. Mowery's analysis did not include the benefits to the larger economy—through jobs generated by strong export performance, for instance—in his analysis. However, despite the fact that job generation is often the framing normative rubric for government support, as it is in the statement above, very few studies attempt retrospectively to account for the record on job generation. In a study of high-tech job generation in the

1970s, we found that high-tech manufacturing industries created only about 20 percent of the jobs lost in other sectors of manufacturing during the decade (Markusen, Hall, and Glasmeier 1986).

Any full accounting of job generation would have to assess the opportunity costs of job creation through other uses of public funds, which are not addressed even in the conceptual arguments. It would also have to assess the negative impact on employment in other sectors through the crowding-out effect—R & D concentrated in some sectors draws scientists and engineers, management expertise, and capital into those at the expense of others. The depletionists cited previously argue that concentrated defense spending in certain sectors has weakened the competitiveness of U.S. civilian industries. Mowery (1981) concludes that federal funding in the postwar period did displace research activity away from sectors receiving little or no federal support, favoring the instruments, electrical machinery, and transportation equipment industries. Our comparison of industry performance in the preceding tables suggest that this has indeed been the case. Should new industries and technologies be underwritten if accelerated displacement of workers in other industries is one of the by-products? Should large, relatively mature industries such as autos or steel be considered critical industries on the basis that they provide many jobs and that the retention of those jobs is in the national interest?

In lieu of projections of job creation potential or other indicators of "success," proposals for choosing critical technologies and industries rely on panels of experts whose ranks are heavily skewed toward representatives of the very industries and firms (and universities and research labs) who stand to gain from the choices made. The proposals would replicate practices that for years have governed the way that advanced military projects are chosen in the Pentagon—where dozens of advisory panels of military officers, industrialists, and academics meet to discuss future research and procurement needs in a relatively incestuous process (Adams 1981). Technology identification and subsidization could easily become an elite-dominated activity and might degenerate into a pork-barrel activity where the more high-tech industries benefit from continual government patronage, divorced from the longer term interests of the public and, arguably, of the economy, too.[6]

A second normative issue is raised by the quote, one related to this last point about governance. What performance criteria should be applied to technology projects, and what returns should the public sector expect for its investment? The second sentence of the quote suggests that making American companies competitive internationally is tantamount to creating large numbers of good jobs for American workers. But the inference is by no means uncontroversial. Making American companies strong does not necessarily guarantee a significant expansion of jobs at home, since new jobs can be located overseas. Growing trends toward the international portability of technology, the multinational character of business, the proliferation of joint ventures and incentives to sell or license technology rather than products abroad, all suggest that the returns to American publicly-financed technology initiatives could easily end up generating employment outside our borders (Mowery and Rosenberg 1989). Companies are vigorously opposing the U.S. preference clauses mandated by Congress in the Cooperative Research and Development Agreements being negotiated in research partnerships between NIST or DOE and private companies, suggesting that this requirement does indeed create a serious constraint on their locational choice.[7]

Currently, many of the technology transfer arrangements commit public sector funding as essentially a gift without any return required on the taxpayer's dollar either in terms of jobs or revenues. An alternative approach to the U.S. preference clause would be to treat public sector technology subsidies as an equity investment, requiring paybacks into the public treasury if they are successful. Several states currently do this with state economic development funds. For instance, Connecticut, in return for royalties, has given several million dollars each to Kaman Corporation and United Technologies to develop commercial applications of defense technologies which if successful will more than recompensate the state for its initial investment. Proposals for government retention of some proprietary rights over new technologies, or at least the right to repossess them if companies do not put them into production, have been discussed, but have not generally been put into practice.

With the elite-controlled and permissive structure currently proposed for new technology initiatives and the absence of performance requirements, the potential for corruption, massive failure, and internationalization of the results, to the prejudice of the American economy, is rather high. Technology policy is being marketed to the American public as a long-term jobs creation program, and rather large sums of money have recently been appropriated for the dozens of categorical programs designed to carry it out. The public debate has been boiled down to a simple "should we, or shouldn't we," framed as a debate between, on the one hand, neoclassical economists and Reagan/Bush era policymakers, who think the market can do fine on its own, and, on the other hand, supporters of the amalgam of dual use and critical technologies, whether inside or outside of the Pentagon. A much more sophisticated and broadly based debate will be needed to ensure that initiatives, which we surely must have, achieve the joint goals of creating socially useful technologies and engender new industries that will provide good jobs for growing numbers of Americans.

The Macroeconomic and Dynamic Dimensions of Technology Policy

A remarkable feature of the current debate about technology policy is the absence of analysis linking it to macroeconomic and long-term growth dynamics. Although it is cast as a long-term strategy for the economy, advocacy of technology policy is based upon microeconomic foundations, on a narrative about competition among nations as analogous to competition among firms. Lester Thurow's (1992) book, for instance, was advertised under the rubric "Who Will Win the Twenty-First Century?," the competitors being the United States, Japan, and Europe. Based on microeconomic theories of the firm and international trade, the prescription for an economy like the United States, is to focus upon the least mobile of its productive factors—infrastructure and the work force (Thurow 1992; Reich 1990). Support for government inducements to technological innovation is included in the formula, because it will help domestic firms win both by introducing new products and by improving technological processes.

There is no guarantee that a world in which the richest industrial nations engage in forced savings to propel technological change forward will necessarily experience increased productivity, growth, and prosperity. Institutionalized state-led innovation could end up being destabilizing and even growth-stunting. First, there is the trade-off between other forms of consumption and investment, and technology investments. Technology investments, planned to be in the tens of billions of dollars a year in the United States, are being paid for out of a combination of higher taxes and lower defense spending. Such spending can crowd out other forms of investment and consumption, with distinct consequences for the macroeconomy and long-term growth rates.

The crowding-out effect is not simply a question of what happens within the technology-oriented nation. Much of the world today is without the benefit of many of the technological achievements enjoyed by residents of advanced industrial countries. What their residents need is technology diffusion that will increase productivity and raise the standard of living. At present, clear choices are being made in the United States to fund speculative technology investments rather than helping Eastern European, former Soviet, African, and South American countries revitalize their economies. If such commitments were to be made instead, orders for machinery and equipment would stimulate the capital goods sectors in the United States, just as higher incomes and wages associated with industrialization in those countries would increase the demand for agricultural and sophisticated consumer goods.

Second, investments in technology will clearly contribute to a worsening of the domestic income distribution in the short and intermediate term. R&D

activities are disproportionate employers of well-educated scientists and engineers, in government labs, private industry, universities, and the nonprofit sector. As defense spending is cut, the ranks of the unemployed and lower income service employees will swell, since the layoffs will fall most heavily among enlisted men and women and blue-collar workers in defense plants. At the same time, the emphasis on higher tech weaponry, dual-use technologies and the conversion of the national weapons labs to environmental and other challenges will protect and in fact increase the ranks of higher income professional and technical workers. This alternation in the income distribution has macroeconomic consequences as well.

Finally, there is the speculative question of the digestibility of rapid technological change in the economy. Long-run growth theorists have in the past been concerned with questions of imbalance in consumption and investment patterns, arguing that it is possible for one to be excessive at the expense of the other. It is conceivable that very rapid rates of technological change could end up being destabilizing. If entire industries are thrown into obsolescence by the accelerated development of new products and technologies, the associated decline of personal incomes of those displaced and mounting social costs could dampen demand and place severe pressures on the state. The economy could be stretched between a large sector of obsolete industries where the costs of resource reallocation mount and a smaller high-tech sector that continues, with government support, to be preoccupied with beating out its international competitors. The possibility could be mirrored in the world at large, where sluggish growth and continued impoverishment continue to constrain international market expansion, creating geopolitical tensions and, of course, limiting demand for the new technologies so deliberately crafted.

This is, of course, a very speculative argument, and one that requires much more thought and research. Nevertheless, the point is that almost no inquiry has been conducted into the linkages between the consequences of different technology strategies and the growth and character of domestic and international markets. Although new technologies and products presumably increase productivity and therefore raise standards of living, the pace at and manner in which they are introduced, and the costs of generating them, might deform the growth trajectory of nations.

Conclusion

Explicit technology policy, as a newcomer to Washington, bears with it a number of unresolved issues. In this paper I have argued that technology policy will not work in the absence of a fuller set of industrial policies, including attention to the issue of market creation in the initial stages. I have also argued that because technology initiatives are risky and expensive, they should be justified in the public interest and accompanied by verifiable performance requirements. Advocacy of technology policy rests on debatable assertions about its payoffs, since little research has been done evaluating the success and failure of past efforts. It is also vulnerable to capture by a narrow spectrum of interests, especially in its current institutional form.

One route out of this dilemma is to harness technology policy to new missions in the economy, which are addressed to pressing national needs, just as past technologies have emerged from the defense mission. Commitments to solving problems on the environmental, energy, and transportation fronts, where the private sector cannot be relied upon, would clarify the public interest issues and provide automatic performance standards. In this approach, public R&D investments would be linked to an ensemble of other initiatives—regulatory, procurement, trade, and financial policies—working together to achieve each mission. We have argued the virtues of taking this path at length elsewhere (Markusen and Yudken 1992, chap. 9). One of its virtues is the relative decentralization of technology initiatives within the federal government, rather than concentration of such resources within DARPA or a civilian technology agency. Furthermore, more parties would automatically be involved in the choice of technique, since the line agencies (Energy, Transportation, EPA) that would be responsible for such initiatives already have mechanisms for including many constituencies in the design and oversight of their policies.

Perhaps the most important conclusion of this rather speculative essay is that no irreversible decisions regarding the size and institutional form of technology policy should be undertaken at this point, without more public debate about the options and better evaluation of the past record. To expand DARPA now, to lavish more funds on the national labs, to create a new civilian technology agency all seem premature—likely to squander precious public sector dollars and set in place an infrastructure that may prove inflexible, politically protected, and difficult to alter as our understanding of advanced industrial economies grows.

NOTES

1. Mowery and Rosenberg (1989) review these features with ample data and well-chosen historical examples. Their work is critical of the narrow analytical inquiry of neoclassical economics that stresses the appropriability phenomenon, and they emphasize instead circumstances influencing firm receptivity to adoption of new technologies, rather than their pursuit through research and development.

2. For the history of these relationships in the aircraft industry, see Stekler 1965; Simonson 1968; Mingos 1968; Freudenthal 1968; Phillips 1971; Roland 1985; Rae 1968; Markusen, Hall, Campbell and Deitrick, 1991; chap. 3.

3. See Flamm 1987; Flamm 1988; 16, 33; Levin 1982; 26; Borrus and Millstein, 1980; 24.

4. These shares are computed from Henry and Oliver 1987 and unpublished data from the Bureau of Industrial Economics. The figures are for military demand only, narrowly construed as orders from the U.S. Department of Defense. This share would have been approximately 75 percent if other U.S. government demand (from NASA) and foreign military sales were included.

5. See also Melman 1974, 1983; Dumas 1986; Ullmann 1989.

6. See also the discussion of this problem in Alic, et al. 1992; 374–77.

7. The OTA (1993) study grapples with the U.S. preference issue, without coming to a definitive conclusion. The worry is that the imposition of the domestic jobs requirement would discourage companies from entering into partnerships with the DOE Labs and other government research facilities that are shopping for a mission to justify their existence and are looking toward private sector partnerships for joint research and technology transfer as a way of surviving. In a landmark agreement between DOE and CSPP, a compromise was reached that requires only that the research and development associated with the CRADA take place in the U.S. Subsequently, other companies have been demanding the same compromise, which is really not a compromise at all on the jobs issue, but unencumbered permission to transfer technology and production (of it and with it) anywhere in the world.

REFERENCES

Adams, Gordon. 1981. *The Iron Triangle: The Politics of Defense Contracting.* New York: Council on Economic Priorities.

Alic, John, Lewis Branscomb, Harvey Brooks, Ashton Carter, and Gerald Epstein. 1992. *Beyond Spinoff: Military and Commercial Technologies in a Changing World.* Cambridge, Mass.: Harvard Business School Press.

Borrus, Michael, and James Millstein. 1980. Trade and Development in the Integrated Circuit Industry. In *American Industry in International Competition,* ed. Laura Tyson and John Zysman. Berkeley: University of California Press.

Dumas, Lloyd. 1986. *The Overburdened Economy.* Berkeley: University of California Press.

Flamm, Kenneth. 1987. *Targeting the Computer.* Washington, D.C.: The Brookings Institution.

———. 1988. *Creating the Computer: Government, Industry, and High Technology.* Washington, D.C.: The Brookings Institution.

Freudenthal, Elsbeth. 1968. The Aviation Business in the 1930s. In *The History of the American Aircraft Industry: An Anthology,* ed. G. R. Simonson. Cambridge, Mass.: MIT Press.

Henry, David, and Richard Oliver. 1987. The Defense Buildup, 1977–85: Effects on Production and Employment. *Monthly Labor Review* (August): 3–11.

Levin, Richard C. 1982. The Semiconductor Industry. In *Government and Technical Progress: A Cross-Industry Analysis,* ed. Richard R. Nelson. New York: Pergamon Press.

Lichtenberg, Frank. 1987. The Effect of Government Funding on Private Industrial Research and Development: A Re-assessment. *Journal of Industrial Economics* 36:97–104.

Markusen, Ann R. 1985. *Profit Cycles, Oligopoly, and Regional Development.* Cambridge: MIT Press.

———. 1991. The Military Industrial Divide: Cold War Transformation of the Economy and the Rise of New Industrial Complexes. *Environment and Planning, D: Society and Space* 9, no. 4.

Markusen, Ann, and Joel Yudken. 1992. *Dismantling the Cold War Economy.* New York: Basic Books.

Markusen, Ann, Peter Hall, and Amy Glasmeier. 1986. *High Tech America: The What, How, Where and Why of the Sunrise Industries.* Boston and London: Allen and Unwin.

Markusen, Ann, Peter Hall, Scott Campbell, and Sabina Deitrick. 1991. *The Rise of the Gunbelt.* New York: Oxford University Press.

Melman, Seymour. 1974. *The Permanent War Economy: American Capitalism in Decline.* New York: Simon and Schuster.

———. 1983. *Profits without Production.* New York: Alfred A. Knopf.

Mensch, Gerhard. 1979. *Stalemate in Technology.* Cambridge, Mass.: Ballinger.

Mingos, Howard. 1968. The Rise of the Aircraft Industry. In *The History of the American Aircraft Industry: An Anthology,* ed. G. R. Simonson. Cambridge, Mass.: MIT Press.

Mowery, David. 1981. The Emergence and Growth of Industrial Research in American Manufacturing, 1899–1945. Ph.D. dissertation. Stanford University.

———. 1986. Federal Funding of Research and Development in Transportation: The Case of Aviation. Paper presented at the National Academy of Sciences Conference on the Effects of Federal R&D, May. Washington, D.C.

Mowery, David, and Nathan Rosenberg. 1989. *Technology and the Pursuit of Economic Growth.* Cambridge: Cambridge University Press.

Nimroody, Rosy, William Hartung, and Paul Grenier. 1988. *Star Wars Spin-offs: Blueprint for a High-Tech America?* New York: Council on Economic Priorities.

Office of Technology Assessment (OTA). 1993. Defense Conversion: Redirecting R&D, OTA-ITE-552, May. Washington, D.C.: U.S. Government Printing Office.

Phillips, Almarin. 1971. *Technology and Market Structure: A Study of the Aircraft Industry.* Lexington, Mass.: Heath Lexington Books.

Rae, John. 1968. *Climb to Greatness: the American Aircraft Industry, 1920–1960.* Cambridge, Mass.; MIT Press.

Reich, Robert. 1990. Who is Us? *Harvard Business Review* 90, no. 1 (January–February): 53–64.

Reppy, Judith. 1989. Technology Flows Between the Military and Civilian Sectors: Theory and Evidence. Paper prepared for the Panel on Defense Spending as Technology Policy for the U.S. Annual Meeting of the Association for the Advancement of Science, January 19.

Roland, Alex. 1985. *Model Research: The National Advisory Committee for Aeronautics, 1915–1958.* Washington, D.C.: U.S. Government Printing Office.

Samuels, R., and B. Whipple. 1989. Defense Production and Industrial Development:

The Case of Japanese Aircraft. In *Politics and Productivity: The Real Story of Why Japan Works,* ed. Chalmers Johnson, Laura D'Andrea Tyson, and John Zysman. New York: Ballinger Publishing.

Simonson, G. R., ed. 1968. *The History of the American Aircraft Industry: An Anthology.* Cambridge, Mass.: MIT Press.

Stekler, Herman O. 1965. *The Structure and Performance of the Aerospace Industry.* Berkeley and Los Angeles: University of California Press.

Stowsky, Jay. 1986. Beating Our Plowshares into Double-Edged Swords: Assessing the Impact of Pentagon Policies on the Commercialization of Advanced Technologies. Working Paper no. 17, Berkeley Roundtable on the International Economy, University of California at Berkeley.

Thurow, Lester C. 1992. *Head to Head: The Coming Economic Battle among Japan, Europe, and America.* New York: Morrow.

Ullmann, John. 1989. Economic Conversion: Indispensable for America's Economic Recovery. Briefing Paper 3, April. Washington, D.C.: National Commission for Economic Conversion and Disarmament.

U.S. Congress, Office of Technology Assessment. 1988. *Paying the Bill: Manufacturing and America's Trade Deficit.* Washington, D.C.: Government Printing Office.

Utterback, James. 1987. Innovation and Industrial Evolution in Manufacturing Industries. In *Technology and Global Industry: Companies and Nations in the World Economy,* ed. Bruce Guile and Harvey Brooks 16–64. Washington, D.C.: National Academy Press.

Vernon, Raymond. 1966. International Investment and International Trade in the Product Cycle. *Quarterly Journal of Economics* 80, no. 2: 190–207.

CHAPTER 7

Does the United States Need a Technology Policy?

W. Edward Steinmueller

The unchallenged U.S. position of industrial and technological leadership in the three decades following World War II has dissipated over the past two decades, leaving U.S. firms to make their way in a world where foreign competitors have the capital, human resources, and government support necessary to prevail in competitive struggles with U.S. industry. What role, if any, might technology policy play in slowing or reversing the decline of American industrial leadership? The topic of technology policy is not new, having been one of the key elements in industrial policy debates throughout the Carter, Reagan, Bush, and Clinton administrations.[1] This earlier debate has shaped many of the terms of reference under which such policy debates are now conducted. For example, criteria for public support of research now generally embrace the ideas that projects should be precompetitive (avoiding the hazards of attempting to pick market winners among competing technological implementations) and generic (suggesting that the spillovers from the technology should be widespread).[2] Exceptions, such as the Clinton administration's Flat Panel Display Initiative and parts of the High Performance Computing Initiative, continue to be justified primarily in terms of national security rather than industrial competitiveness. The approach adopted here is to examine the underlying rationale of such policies, revisiting the history and principles under which technology policy discussions have been framed. As such, this discussion is a top-down discussion of technology policy[3] that largely bypasses the major problems that have been experienced in implementation of the policy and that are considered in bottom-up investigations of specific policies and the industries. These studies reach pessimistic assessments of the value of pursuing technology policy by induction.[4]

From an international perspective, the recent U.S. debates on industrial policy appear to be an unusually vigorous exercise of ideological fervor in promoting laissez-faire models of economic organization. In contrast to the highly pessimistic conclusions of studies concerned with industrial or technology policy in the United States, it has been argued that different structures of partnership relationships between government and industry in developed nations and promotional or paternalistic interventions by government in rapidly developing nations have played an important role in catching up to the technological performance of the U.S. economy.[5] One means of resolving this apparent contradiction is to argue that the tradition of industrial and technology policy in the United States has been *dirigiste* (government-directed) and that technology policies in particular have failed due to the absence of industrial participation in framing the policy.[6] Another is to demonstrate that the coexistence of proactive foreign policies and technological performance is largely coincidental, discounting the causal relationship between foreign government policies and the success of foreign enterprises in overtaking U.S. enterprises.[7] How this debate is resolved will influence future industrial and technology policy debates.

To address the distinct issue of technology policy it is necessary to ascertain a boundary between industrial and technology policy. The range of possible industrial promotion policies that may have an impact on technological development or industrial competitiveness is broad, encompassing fiscal policies in taxation and government procurement, legal and regulatory decisions promoting domestic over foreign business interests, foreign trade policies for preserving traditional or promoting infant industries, and non-technological infrastructural investments that serve domestic business. These industrial policies may either be complements or substitutes for specific technology policies.

The definition of technology policy is generally confined to a much narrower range of issues involving preferential tax treatment of research-intensive industry investments, direct and targeted public investments in research and development, and procurement or contracting procedures that seek to accelerate or shape the course of technological development. This definition of technology policy places excessive emphasis on the supply side and on research-intensive industries, thereby diverting attention from the demand for and use of technologies. It also ignores the role of complementary actions, such as the distribution of knowledge or the design of organizational change, that may or may not be technologically related.[8] In addition, between industrial policies and explicit, narrowly constructed technology policies there is a gray area in which public policies in education, health, and environmental policies may either favor or discourage technologically progressive developments depending upon their content. Technology policy should be aimed at "technologically progressive" industries, those that either produce or bring into use new technologies that enhance productivity or stimulate economic growth. Thus, a broad view of technology policy recognizes the role of demand (both direct and derived), complements, and supply-side factors in shaping technological developments as well as the use of technology, and it interacts directly and actively with industrial policy considerations.

Although there is much to learn from the shortcomings of past initiatives, proponents of new initiatives are likely to regard their cause as unique. It is therefore useful to examine the mechanisms by which past policies have had a positive influence on technological development. The principal theme of this chapter is that the United States has adopted policies in the past that have had a salutary effect on both technological and economic performance. Often the greatest success has occurred with policies whose effect on technological development was unintended or incidental. Moreover, it would be unrealistic to expect an end to policies of technological promotion favoring specific industries in a pluralistic democracy. Technology policies will continue to proliferate.

Finally, in this chapter competitiveness is not the same as productivity advance.[9] An aim of technology is to retain and expand technologically progressive industries because these industries are more likely to contribute to future productivity gains and economic growth than those industries where technological developments are absent or moribund. This focus leaves room for industrial policies focused on non–technologically progressive industries that coincidentally make major contributions to productivity growth or to the conditions supporting productivity growth more generally. A few remarks on the perceived relative decline of U.S. industrial and technological leadership will set the stage for examining past and future technology policy initiatives.

In examining the origins of the relative U.S. decline two factors are worth emphasizing, the immense costs of the Cold War and the difficulties in maintaining leadership given the convergence of technological capabilities of industrialized nations. In a very real sense, the United States has recently emerged from a half century of military struggle whose cost, fortunately, has predominantly been measured in economic resources rather than human lives. If we assume that the Cold War elevated defense expenditures by one-half over what they might have otherwise been, the U.S. defense budget has for nearly 50 years carried an "excess third" unmatched by our industrial rivals. This excess third of military expenditures from U.S. investment in the Cold War amounts to having fought two more World War IIs in the postwar era. From a technological viewpoint, the Cold War diverted approximately 30 percent of U.S. scientific and technical resources for the past half century. Even if one takes an optimistic view of defense spillovers to the civilian sector, the economic benefits to the private sector from the excess third of military expenditures, concentrated in the deployment of U.S. force to meet the Soviet threat and the spiraling costs of nuclear deterrence, are where we would expect the fewest civilian spillovers. The cumulative effects of the Cold War investment are adequate, alone, to explain much of the relative decline in U.S. industrial leadership over the past half century given the fact that our industrial competitors have spent, at most, two-thirds as much as a proportion of GNP on defense as we have. In the case of Japan, in particular, the expenditure as a proportion of GNP has been less than one-fifth that of the United States.

Although the costs of the Cold War diverted enormous resources from private consumption and investment over a prolonged period, the relative decline in U.S. leadership is often dated from the beginning of the first oil shocks of the 1970s and the relative decline in U.S. technological leadership from the end of that decade. Neglecting the real possibility that the relative decline in this period was the result of the cumulative effect of the excess burden of U.S. defense costs in reducing human and physical capital formation and diverting public sector investment, some decline in American leadership must have been expected with the convergence in output and productivity among industrialized nations.

The convergence hypothesis has been extensively examined in an enormous literature.[10] For our purposes, two elements of a much more complex account will suffice. The first point, examined by Abramovitz (1986), is that economic leadership provides a target for imitation by which the businesses of other nations may structure their activities in terms of a catch-up process. By this account, American leadership following the devastation of World War II was a natural target that served to define the growth potential of other nations and to support a relatively rapid rate of productivity and output growth. By contrast, the problem of forging ahead, involving the exploration of untested technologies and the construction of new industries, has been relatively more difficult, accounting for a lower long-term increase in output and productivity of the leader, the United States. Thus, convergence emerges as the result of the relative disparity between the growth potential of nations engaged in catching up and the problem of the leader in forging ahead.

While the Abramovitz explanation is consistent with the aggregate patterns of long-term output and productivity growth, it does not explain the breadth or extent of technological advantage that the United States was able to retain for over 25 years after the end of World War II. Nelson (1990) argues that the immediate postwar era is marked by an extraordinary set of public policy decisions that enhanced American abilities to forge ahead and develop virtually uncontested leadership in many of the industries at the technological frontier. The three factors he notes are (1) the dramatic increase in the percentage of American high school graduates that went on to college as the result of American postwar affluence and, most significantly, as the result of government support for student enrollment under the GI bill of rights and other public fellowship programs for college attendance; (2) massive funding for university research that, for many years, was not matched by Europe or Japan and that propelled U.S. higher education to a historically unprecedented international position; and (3) the development of enormous applied research efforts by the Department of Defense and later by the National Aeronautics and Space Administration (NASA) that had no parallels among other capitalist economies and that, in their early years, created large and unanticipated spillovers to the civilian sector.[11]

The government policies that Nelson identifies as responsible for the U.S. breakaway in forging ahead of rivals during the first two decades following World War II are an excellent foundation for considering the current prospects for U.S. technology policy. Each of these policies is implicitly linked to a possible source of advantage from technology-related policies. Stated in more abstract terms, Nelson's account offers three possible avenues for gaining technological and industrial leadership through government policy. The first is to forge ahead of rivals in the rate of human capital formation. The second is to lead rivals in developing some important component of the national infrastructure supporting technological progress. The third is to undertake technologically intensive national missions that stimulate the supply of technologies with significant commercial spillovers. I will add a fourth avenue: government policies that are aimed at forging ahead in the creation of new types of demand. This consideration of demand extends Nelson's historical discussion of mission-focused activities of the DOD and NASA in creating technological spillovers to the prospective role of government procurement or other measures that seek to engage technological development through demand-side direct effects and spillovers.

The four principles just described are the basis for the following sections. While these principles identify the avenues by which technology policy may contribute to advancing U.S. leadership, they also present a major set of analytical and practical problems. From an analytical viewpoint, the problem is one of determining if, and specifying where, a productive margin exists for forging ahead of rivals. From a practical viewpoint, pursuing a particular margin of advance requires the adoption of specific policies that may require overcoming substantial political and social barriers. Brief comments on the problems of policy implementation are incorporated in the concluding section.

Simply reproducing the historical record by which the United States rapidly advanced its leadership following World War II is not possible. The potential for forging ahead depends on historical developments; it is path-dependent. What earlier served as a productive margin by which the technological frontier could be advanced may now be subject to diminishing returns or may offer little relative advantage given rivals' capabilities and intents. For example, if commercial spillovers available from military technology to the civilian sector are diminishing as the result of divergence of the two sectors' requirements, military expenditure is foreclosed as a potential avenue for the relative advance of the United States over rivals. Similarly, the advantages available from the American system of mass production and distribution offer fewer opportunities for relative gain given the success of industrial rivals in developing similar systems. The limited life of mechanisms for forging ahead of rivals is a central problem of technological and industrial leadership.

Human Capital Formation

Both the level and composition of human capital investments determine the capabilities of a country to either advance the technological frontier or to bring a large proportion of an industry to the technological frontier. As noted, U.S. technological leadership following World War II was strengthened by extending opportunities for citizens to participate in higher education. These opportunities were further extended during the 1960s when significant improvements in the educational attainment of women were achieved. The United States, however, is still deficient in extending the opportunities for postsecondary education to the African-American and some other minority communities. Nonetheless, American participation in higher education in the 18–22-year age-group has now been matched by most of our major rivals. Moreover, U.S. primary and secondary education appears to have developed significant deficits in performance relative to our industrial rivals. While the United States can, and should, make significant strides in remedying these deficiencies, there appears to be little opportunity for forging ahead of rivals by raising the level of educational attainment.

Greater opportunities exist in altering the composition of human capital formation through the educational system. The United States has emphasized the higher education system as the means for accumulating high levels of generalized knowledge and skill. The rapid and accelerating pace of technological change suggests that this emphasis has been appropriate; generalized abilities at learning and problem solving are the best tools for coping with rapid change. The broad range of skills in learning necessary for completing a liberal or technical college education have increased the flexibility and adaptability of the college educated portion of the U.S. labor force.

Despite the efficacy of college education at developing what Reich (1991) calls "symbolic analyst" skills, there is a significant and growing gap between the skills needed to adequately participate in the labor force and those offered secondary and community college graduates. Appeals for competition policy and a more sophisticated national policy regarding technology have been uniformly vague in defining blueprints for retraining and skill acquisition by the less well educated portion of the U.S. labor force. The most extensive discussion (11 pages) occurs in Dertouzos, Lester, and Solow (1989). They are candid about the shortcoming of the present opportunities offered secondary graduates ("there is no systematic path to training for the non-college-bound" [p. 85]) and about retraining of unemployed workers ("U.S. experience is a dismal record of failure" [p. 91]). Their own proposals amount to an appeal for lifelong education given the certainty of job change and a commendation of formal industrially oriented education as practiced by Sweden, the apprenticeship system of Germany, and the corporate retraining efforts of some of Japan's largest companies (e.g., Nippon Steel). The Cuomo Commission Report (Smith 1988) recommends government incentives for industry-based training programs. Piore and Sabel (1984) argue that such skills can only be transmitted through the resurgence of a craft system in which the arm's-length relation among business and other social institutions is dissolved. Cohen and Zysman (1987) argue that improving skill levels is an urgent priority for restoring American manufacturing, but are vague about what skills should be emphasized or how training should be delivered. More detailed examinations of the changing skills of workers within the U.S. labor force have uniformly emphasized uncertainty as a key characteristic in relating technological change to the nature of required skills.[12]

Current American capabilities for providing entry level skills and for retraining of workers over a lifetime of frequent employment changes have clear shortcomings. Aside from the problems that this shortcoming creates for individuals, is this problem a significant issue in the ability of the United States to restore its technological leadership?

Technological leadership may be defined in terms of the leading or cutting edge of technological advance. As noted earlier, leading nations face substantial problems in maintaining the pace of advance at the frontier. It is also true, however, that the benefits from technological advance can only be realized by adoption of leading-edge technologies. Economists have historically placed too much emphasis on the assumption that technological information could be reproduced and distributed at low marginal costs. While there are instances where this is true, much of the knowledge that is useful for technological and manufacturing leadership is localized and tacit. Knowing that a blast furnace is a high temperature chemical reactor is very different from being able to tell when to pour steel. Correspondingly, the worker who is only taught a procedure for inputting information into an information system cannot be expected to provide meaningful feedback about the failure of the system to provide the expected result. The broad adoption of new technologies requires a knowledgeable and critical labor force. We need a better statement than is presently available about the skills that such a labor force needs to have in the American context, and we need to devise methods to deliver such skills. Developing such skills should be a leading priority of technology policy since it offers the best hope that further advances in the frontier of technology can be broadly utilized.

Technological Infrastructure

Recent technology policy discussions have centered on the issue of how government policy might enhance the technological infrastructure that business enterprises can utilize to grow and prosper. A major component of the U.S. infrastructure has been the research university, and there is evidence that investments in university research continue to offer substantial returns to the business sector.[13] Despite the continued strength of the U.S. research university, there have been recent calls for universities to engage in a more direct partnership with business firms in the development and commercialization of technology. For example, while the Young Commission endorsed the principle that universities should continue to be the primary performers of basic research, they also concluded that universities should undertake "to focus research on problems that industry sees relevant to marketplace needs."[14] The common perception that little action has occurred on this margin is incorrect. Substantial changes in the role of universities is under way, and many of these changes are the direct result of federal policy initiatives.

Universities increasingly are regarded as potential sources of industry-relevant research.[15] Science and technology parks sited in the immediate environs of research university campuses are by now ubiquitous features of the landscape and have become important components in economic development plans. During the 1980s some 286 university-industry research centers were formed in the United States, more such centers than had been established in the country throughout the preceding century (1880–1980).[16] These developments were strongly encouraged by federal government priorities. Federally funded R&D centers themselves enjoyed a tripling in the level of their support (in real dollar terms) between 1975 and 1990. Active participation of industrial sponsors in interdisciplinary, university-based science and engineering centers has been solicited with the encouragement of the National Science Foundation.

The recognition that universities are sources of industry-relevant research has been heightened by the growing ties between basic and applied research that are a concomitant of the increasing dependence of many industrial technologies on scientific theories and theoretically informed technological developments. Moreover, the character of industrial research activity itself has been shaped by modern methods of scientific investigation, becoming more theory-based, model-driven, and deductive than earlier methods of empirical search and inductive reasoning. This development is particularly evident in pharmaceutical research, where testing of natural and synthesized chemical samples has given way to molecular design based on physiological models.[17] While the primary purpose of university performed basic research continues to be the pursuit of scientific and engineering knowledge according to an internally generated and validated agenda, the conduct of basic research in this setting often yields by-products of importance to society at large.

Reconciling the management of the university research enterprise as an educational institution, as the performer of basic research, and as a major element in the technological infrastructure for U.S. industry is by no means an easy problem.[18] The United States is pioneering this complex relationship; recent developments suggest that barriers that previously separated the university from industry are rapidly disappearing, and the effects of these developments on the university as an institution remain uncertain.

Non–university-based research initiatives aimed at industrial technology leadership in the United States remain highly focused on the industries that have traditionally played a central role in national defense. The extensive development of national laboratories during the Cold War has resulted in an enormous research capability that will be difficult to harness for commercially relevant purposes. As Markusen (1999) notes, the majority (54 percent) of federal spending for the industrial performance of R&D is devoted to the aerospace industry while another 27 percent flows to the electronics and communications industries. While this sectoral imbalance in research funding may partially reflect differences in "technological opportunities," it is worth considering what constitutes cause and effect of such opportunities. If national defense had required substantial augmentation of the U.S. natural resource base, would the agricultural and energy sectors be a more important source of commercial innovation?

The relative erosion of American technological leadership has also elicited an industrial response. Private research initiatives involving research consortia funded by industry such as MCC or by government-industry partnerships such as Sematech are examples. While the resurgence of the U.S. semiconductor and semiconductor equipment industry may be partially attributed to Sematech's efforts, the role of individual company innovation and intellectual property protection of U.S. microprocessor designs deserves the lion's share of the credit.[19]

Dramatic initiatives such as building new and large-scale institutions in the research infrastructure or broadening public support for private initiatives such as industry research consortia require caution and careful examination. At present, few capabilities exist for managing such efforts at the federal level, and the tools available for setting these institutions' goals are modest. The two methods for setting goals presently available are (1) the identification of gaps between possible and existing national technological capability or practice and (2) the identification of a specific frontier that can capture the imagination of large segments of the society. In the United States, the latter method has most commonly been organized around nationally funded missions (to be considered in the next section), such as the space effort. With regard to gaps, the United States has a limited experience with indicative planning, a process in which a consensus is achieved about desirable paths of development to close such gaps and which is followed by the mobilization of social resources. Sematech has come closest to following this model in identifying shortcomings in U.S. IC process equipment markets and technology relative to Japanese rivals. It is an open question whether the Sematech model of joint government-industry funding can or should be more broadly adopted in other U.S. industries.[20]

Supply Spillovers

In an era of budgetary stringency it may be inopportune to suggest that a key component of past U.S. technology policy has been the development of new federally financed national missions. Several generations of defense technology including the development of radar, long-range bombers, the Minuteman missile, and nuclear submarines have in their early stages created significant commercial spinoffs—commercial air traffic control, civilian jet aircraft, integrated circuits, and nuclear power technology. Similarly, the space program produced satellite telecommunications. In each of these examples, one may question the necessity of the federal mission in establishing the respective industry, e.g., integrated circuits, or the particular nature of the civilian technology that was created, e.g., nuclear power. Nonetheless, in each case the U.S. achieved and, in most cases, retained technological leadership in these areas.

Unfortunately, recent candidates for new national missions such as energy independence or environmental leadership have proven monumental undertakings fraught with peril as illustrated by the fate of the Carter administration with the former example. American abilities at concentrating federal expenditures have often been defeated by congressional attention to spreading expenditures among constituencies, and recent achievements in defining national technology missions are somewhat discouraging. The space lab, the national aerospace plane, the superconducting supercollider, and the human genome initiative seem to confirm the proposition that the budgetary allocation process works on the principle that bigger is better.[21] Each of these projects has shortcomings in considering the trade-offs in time between advancing technology and the current cost of seizing a leadership position. In this respect, the federal government appears to want to model efforts on the Manhattan Project rather than its own experience with agricultural technology, satellite telecommunications, or computer technology.

Many of the opportunities for forging ahead of industrial rivals represent extremely complex combinations of institutional, social, and technological innovation. Many of these problems, including enhancing environmental technology, improving productivity in the American education system, and accelerating the pace of manufacturing automation, only partially depend upon the rate of improvement of the hardware that might aid in their solution. There are, however, several opportunities for new national missions including rapid development of the communication infrastructure or application of biotechnology to agriculture and medicine. In each of these cases, as well as other more specialized endeavors, it is important to avoid a "command and control" approach in which an agency is charged with the responsibility for finding a single best solution and implementing it. The only hope of success for such initiatives is through building government-industry links that are open-ended to method, receptive to experimentation, and committed to outcome.

Industrial and technology policy have an important difference. Industrial policies must focus on criteria like increasing employment, achieving near-term improvement of growth rates, and altering the composition of international trade. Industrial policy sets policy priorities for new national missions by identifying industrial targets, and technology policy may be relegated to a secondary position, one in which industrial policy creates a derived demand for technology policy. This ordering is enormously risky since its potential success crucially depends on correctly specifying the industrial policy target. By contrast, a technology policy based on indicative planning and modified by some attention to the comparative abilities to develop the results offers a much broader frontier for the market system to select winners. In addition, investments in technology policy are appropriately leveraged. Much more modest investments are required to advance technologies to the development stage in a number of industries than to attempt to assure the international success of any one industry.

Demand Externalities

The dominance of military procurement in federal government expenditure suggests at first glance that the federal government has, since the completion of the interstate highway system, had little ability to affect the nature of demand in the United States. A closer examination, however, suggests that the major demand impacts of government policy during the past 20 years have been in the area of regulation. Accompanying the deregulation of the transport and communications sector, the United States has experienced a major growth in social and environmental regulation. For example, the costs of environmental regulation increased from about $30 billion in 1972 to about $140 billion in 1990.[22] Much of this expenditure has involved the investments in abatement technologies and new, environmentally cleaner, production technology. Similarly, U.S. policies regarding health-care reimbursement have had a powerful impact in shaping the demand for advanced diagnostic equipment and pharmaceutical products. In these and many other cases, the demand implications have had a major impact on technological innovation and diffusion. Examinations of the impact of regulations have, unfortunately, been limited to considerations of the effectiveness of such programs in meeting stated goals or computation of costs and benefits narrowly defined.

The unintended consequences of regulation have been large and are likely to continue to expand. Economists have played a growing role in suggesting more efficient methods of realizing these social objectives. At present, however, there is little consideration of how regulatory costs shape the pattern of demand for innovation and new technology or how alternative policies would produce different outcomes in technological progress. This outcome is particularly notable in the environment sector where "best available practice" policies have created an international sweepstakes for abatement equipment. The failure to consider the effects of regulation on the derived demand for new technologies is one of the more convincing arguments for the need to expand U.S. technology policy analysis efforts. The current situation, in which the Office of Technology Assessment has been dismantled and the Congressional Budget Office is under budgetary pressure is particularly unfortunate in this regard. Devoting federal resources to the analysis of the consequences of federal policy on technology is a simple recognition that the federal government continues to have a major impact on technological development and is independent of the more controversial undertakings of creating an indicative analysis, forecasting, or planning apparatus within the federal government.

The federal government may yet have an important direct role in demand through efforts to stimulate research and educational productivity. The importance of information technology in both the service and manufacturing sectors is undeniable. For example, depending on how the National Research and Education Network initiative is implemented, the federal government may play a major role in funding research on the use of data communication networks, in defining standards for the transmission of information, and in procuring or mandating standardized components for implementing such a system. While there are ample opportunities to make major blunders in the implementation of the information superhighway, it is doubtful that the market will deliver the universal access that was the basis of network externalities in our current voice telecommunications system. At the same time, there is no reason that a national data communications highway needs to be built by the federal government or even needs to have a common structure. The existing Internet network is built on heterogeneous hardware foundations. Here, as in the case with new national mission initiatives, the temptation to believe that big is better should be avoided.

Implementation and Conclusion

The reserved optimism of the preceding remarks on the prospects for technology policy must be tempered by two closing remarks. First, the United States is not alone in efforts to advance the technological frontier in areas that might expand output growth potential, nor is the United States equally well positioned to take advantage in all lines of technological advance. Recent discussion of HDTV strongly emphasized the difficulty of U.S. companies reentering the consumer electronics market where advantage long ago passed to Japan. At the same time, most scholars and business people that have closely examined technologically progressive sectors of American industry have noted the erosion in the U.S. infrastructure supplying materials, instruments, and specialized production equipment. The faith in the responsiveness of markets in immediately creating these elements of industrial infrastructure when needed has been the shield by which decline has been ignored. This faith has not yielded to the growing body of evidence that there are substantial learning economies, accumulations of localized knowledge, and network externalities in technologically progressive industries. Nor has the policy debate considered in other than a superficial way how other countries manage their technology efforts. In particular, the United States has recast European and Japanese efforts in terms of our own experience with mission-oriented research. For example, Japan's VLSI program is often cited as an example of coordinated effort among cooperative players when it was in many important respects an effort to achieve a threshold and focused effort among highly competitive and uncooperative corporations.[23] Improving technology policy will require a much broader and deeper effort to understand how technology comes to be utilized in advanced economies, how interdependencies develop across firm and industry boundaries, and why investments in technological improvement are focused in a relatively few industries and differ substantially among firms even within those industries.

Second, while I have noted considerable shortcomings in our current abilities to define goals for federal technology policy or even to evaluate the unintended effects on technological progress of existing federal policies, there is a widespread suspicion that even with these capabilities, the federal government would be ineffective in implementing the policies that might be effective. In this area it is not enough to say that insufficient resources have been expended in the analysis of past experience, although this is true. Nor is the problem simply one of oversight. Past failures of technology policy have often been candidly assessed by Congress, by the executive branch, and within the academic community. At least three factors are having a corrosive effect on our ability to implement policies successfully. First, the budgetary allocation process of the U.S. government has been captured by the same parties that are responsible for program oversight. As a result, the representatives and senators who are most intimately involved with a given mission agency are the same individuals that develop the agency's budgetary authorization. Trade-offs between different efforts are made only at a very aggregate level. There is no clear path out of this particular thicket, although the development of an independent authority charged with the responsibility for evaluating the overall effectiveness of U.S. technology policy might help. Second, the narrowness of the national missions that create the derived demand for technology policies relative to our historical reliance on the Department of Defense has created a vacuum not only in policy formulation but also in implementation. Again, there is no immediate solution. Creating a large-scale civilian technology agency would be an enormously hazardous undertaking given the insufficient grounding of such an initiative in policy experience and institutional history. Third, we have had a long national experience with the philosophy that government is inherently a defective institution for achieving any worthwhile social purpose. It will require time and contrary evidence for an alternative view to become more broadly accepted.

The main role for technology policy in the United States is to support the advance of the technological frontier that underlies the long-term growth potential of the U.S. economy. The mission-oriented agencies of the U.S. government have developed many technology policies, and the unintended consequences of federal regulatory policies have, by default, produced even more technology policies. It would be impractical and hazardous to centralize decision-making authority in the formulation or execution of technology policies, but it is a worthwhile effort to develop the analytical and empirical resources to examine the effects of existing policies. Restoring American technological leadership or stemming its erosion is a worthwhile economic goal because the long term-growth of the U.S. economy depends upon the creation and diffusion of the new technologies that offer not only the prospect of new industries but also the opportunity to improve the productivity of existing industries and meet increasing global competition. It is, unfortunately, not possible to reproduce the factors responsible for our past technological leadership. So we must reexamine our existing institutions and develop new ones to meet the challenge of creating the complementary public goods necessary for forging ahead of industrial rivals in technology including human resources and technological infrastructure. Our past experience with university research and mission-oriented policies offer unique opportunities for new technology policies differentiated from those of our industrial rivals. To have any hope of success, we must develop a much broader and deeper ability to analyze our current and past technological activities, we must make institutional reforms in the process by which technology policy is formulated and implemented, and we must avoid the temptation to concentrate our efforts on any single or small collection of major initiatives.

NOTES

The author is grateful for financial support from the Center for Economic Policy Research, Stanford University, allowing the original preparation of this chapter.

1. See Graham 1992 for a review of the industrial policy debate in the period preceding the Clinton administration.

2. One recent contribution that observes the current rules regarding discourse in this area is Tassey 1992.

3. A similar plan of investigation is adopted by Ergas (1987).

4. For examples of studies that approach this issue from a bottom-up approach see Nelson and Langlois 1983, Nelson 1984, and Cohen and Noll 1991.

5. In Procassini 1995, the president of a leading U.S. industrial association (the SIA) reflects on the history of relationships among industrial associations and government in Japan, the United States, and Germany. Zysman (1977) has chronicled the French experience with industrial order. Amsden (1989) and Hobday (1995) have described government interventions in the Korean economy. Okimoto (1989) and Johnson (1982) have examined the Japanese experience.

6. For an example of an alternative approach see Romer 1993.

7. See, for example, Fransman 1990.

8. For a detailed discussion of the policy implications of a greater attention to demand-side issues in the diffusion of knowledge see David and Foray 1995, 1996.

9. While there is considerable merit in Krugman's (1994) arguments regarding the potential dangers of an obsession with competitiveness, technologically progressive industries have historically made a disproportionately large contribution to aggregate growth in productivity and output. Neglecting the "competitiveness" of these industries is a gamble that future productivity advances will be distributed evenly throughout the economy.

10. See Gerschenkron 1952; Maddison 1964; Dennison 1967; Baumol, Nelson, and Wolff 1994.

11. Nelson offers four explanations for the decline of leadership: (1) the free flow of international trade has reduced U.S. domestic advantage in serving the world's largest market, the United States; (2) internationalization of business accelerated the flow of technology across national borders; (3) other nations have attained or surpassed U.S. investments in science and engineering education and civilian R&D; and (4) there has been a sharp decline in the importance of spillovers from military research into civilian technology due to the growing divergence between military and civilian technological requirements. The first and second explanations raise questions, about whether the United States has or has not been a net beneficiary of liberalized trade, that are outside the scope of this chapter. The third and fourth are considered below.

12. See Spenner 1988 for a review of the case study literature.

13. See Mansfield 1991.

14. U.S. Office of the President 1985, 71.

15. See David and Steinmueller (forthcoming) for a review of these issues.

16. Cohen, Florida, and Coe 1994.

17. See Arora and Gambardella 1994.

18. See David, Mowery, and Steinmueller (forthcoming).

19. Grindley, Mowery, and Silverman 1996.

20. See Grindley, Mowery, and Silverman 1996 for a similar conclusion.

21. Shapley and Roy (1985) argue this point forcefully.

22. Carline 1991.

23. See Fransman 1990.

REFERENCES

Abramovitz, M. 1986. Catching Up, Forging Ahead, and Falling Behind. *Journal of Economic History* 46:385–406.

Amsden, A. 1989. *Asia's Next Giant: South Korea and Late Industrialization.* New York: Oxford University Press.

Arora, A., and A. Gambardella. 1994. The Changing Technology of Technological Change. *Research Policy* 23:523–32 (special issue in honor of N. Rosenberg).

Baumol, W., R. Nelson, and E. Wolff. 1994. *Convergence of Productivity: Cross-National Studies and Historical Evidence.* New York: Oxford University Press.

Carline, A. 1991. *Environmental Investments: The Cost of a Clean Environment, A Summary.* December. Washington, D.C.: Environmental Protection Agency.

Cohen, L., and R. Noll. 1991. *The Technology Pork Barrel.* Washington, D.C.: Brookings Institution.

Cohen, S., and J. Zysman. 1987. *Manufacturing Matters: The Myth of the Post Industrial Society.* New York: Basic Books.

Cohen, W., and D. Levinthal. 1989. Innovation and Learning: The Two Faces of R&D. *Economic Journal,* 569–96.

Cohen, W., R. Florida, and W. Coe. 1994. University-Industry Research Centers in the United States. Carnegie Mellon University, typescript.

David, P. and D. Foray. 1995. Accessing and Expanding the Science and Technology Knowledge Base. *STI Review* 16: 14–68. Paris: OECD.

———. 1996. Information Distribution and the Growth of Economically Valuable Knowledge: A Rationale for Technological Infrastructure Policies. In *Technological Infrastructure Policy,* ed. Teubal et al., 87–116. Dordrecht: Kluwer.

David, P. A., D. Mowery, and W. E. Steinmueller. Forthcoming. University-Industry Research Collaborations: Managing Missions in Conflict. In *A Productive Tension,* ed. P. A. David and W. E. Steinmueller (forthcoming).

David, P. A., and W. E. Steinmueller. Forthcoming. *A Productive Tension: University-Industry Research Collaborations in the Era of Knowledge-Based Economic Development.* Stanford: Stanford University Press.

Dennison, E. 1967. *Why Growth Rates Differ.* Washington, D.C.: The Brookings Institution.

Dertouzos, M., R. Lester, and R. Solow. 1989. *Made in America: Regaining the Productive Edge.* Cambridge: MIT Press.

Ergas, H. 1987. Does Technology Policy Matter? In *Technology and Global Industry: Companies and Nations in the World Economy,* ed. B. Guile and H. Brooks. Washington, D.C.: National Academy Press.

Fransman, M. 1990. *The Market and Beyond: Information Technology in Japan.* Cambridge: Cambridge University Press.

Gerschenkron, A. 1952. Economic Backwardness in Historical Perspective. In *The Progress of Undeveloped Areas,* ed. Bert F. Hoselitz. Chicago: University of Chicago Press.

Graham, O. Jr. 1992. *Losing Time: The Industrial Policy Debate.* Cambridge: Harvard University Press.

Grindley, P., D. C. Mowery, and B. Silverman. 1996. The Design of High-Technology Consortia: Lessons from Sematech. In *Technological Infrastructure Policy,* ed. M. Teubal et al. Dordrecht: Kluwer.

Hobday, Michael. 1995. *Innovation in East Asia: The Challenge to Japan.* Aldershot, U.K.: Edward Elgar.

Johnson, C. 1982. *MITI and the Japanese Miracle: The Growth of Industrial Policy, 1925–1975.* Stanford: Stanford University Press.

Krugman, P. 1994. Competitiveness: A Dangerous Obsession. *Foreign Affairs* 73, no. 2: 28–44.

Maddison, A. 1964. *Economic Growth in the West.* New York: The Twentieth Century Fund.

Mansfield, E. 1991. Academic Research and Industrial Innovation. *Research Policy* 20:1–12.

Markusen, A. 1999. Can Technology Policy Serve as Industrial Policy? In this volume.
Nelson, R. 1984. *High Technology Policies: A Five Nation Comparison.* Washington, D.C.: American Enterprise Institute.
———. 1990. What Has Happened to U.S. Technological Leadership? In *Technological Competition and Interdependence: The Search for Policy in the United States, West Germany, and Japan.* G. Heiduk and K. Yamamura. Seattle: University of Washington Press.
Nelson, R., and R. Langlois. 1983. Industrial Innovation Policy: Lessons from American History. *Science* (February 13): 814
Okimoto, D. 1989. *Between MITI and the Market: Japanese Industrial Policy for High Technology.* Stanford: Stanford University Press.
Piore, M., and C. Sabel. 1984. *The Second Industrial Divide: Possibilities for Prosperity.* New York: Basic Books.
Procassini, A. 1995. *Competitors in Alliance: Industry Associations, Global Rivalries and Business-Government Relationships.* Westport, Conn.: Quorum Books.
Reich, R. 1991. *The Work of Nations.* New York: Vintage Books, Random House.
Romer, P. 1993. Implementing a National Technology Strategy with Self-Organizing Industry Investment Boards. *Brookings Papers on Economic Activity,* 345–90.
Shapley, D., and R. Roy. 1985. *Lost at the Frontier: U.S. Science and Technology Policy Adrift.* Philadelphia: ISI Press.
Smith, L. 1988. *The Cuomo Commission Report: A New American Formula for a Strong Economy.* New York: Simon and Schuster.
Spenner, K. 1988. Technological Change, Skill Requirements, and Education: The Case for Uncertainty. In *The Impact of Technological Change on Employment and Economic Growth,* R. Cyert and D. Mowery, 131–84. Cambridge: Ballinger.
Tassey, G. 1992. *Technology Infrastructure and Competitive Position.* Norwell, Mass.: Kluwer Academic Publishers.
Teubal, M., D. Foray, M. Justman, and E. Zuscovitch, eds. 1996. *Technological Infrastructure Policy: An International Perspective.* Dordrecht: Kluwer Academic Publishers.
U.S. Office of the President. 1985. President's Commission on Industrial Competitiveness. *Global Competition: The New Reality,* vol. 2, John Young, Chairman. Washington, D.C.: USGPO.
Zysman, J. 1977. *Political Strategies for Industrial Order: State, Market, and Industry in France.* Berkeley: University of California Press.

PART 4

Competitiveness and Industrial Policy

CHAPTER 8

A High-Road Policy for U.S. Manufacturing

Daniel Luria

Introduction: A New World for Manufacturing

A series of epoch-making (or, more accurately, epoch-ending) events in the early 1970s changed U.S. manufacturing forever. The freeing of the dollar from gold in 1971 and the sudden (albeit temporary) jump in the prices of oil and other energy feedstocks are the macroeconomic highlights, but not the decisive changes for companies that made manufactured goods in the United States. For them, neither exchange rates nor energy prices had anything like the effect of the qualitative collapse of predictable product demand. In 1973, virtually every U.S. manufacturing industry—defined at the two-, three-, and even four-digit Standard Industrial Classification level—had record-high shipments (sales), value added, employment, payroll, and payroll per employee (U.S. DOC 1967, 1972, 1973).

In 1974 and 1975, the U.S. economy suffered an acute recession, the first such recession since 1957–58. As the economy recovered in 1976–78, many industry analysts assumed that the standard postwar business cycle was intact. And, in their defense, many U.S. manufacturing sectors had returned to health, some exceeding by 1978 or 1979 their prerecession peak performance levels (U.S. DOC 1967, 1972, 1973, 1978, 1979).

But many had not or, if they had, they'd done so by changing course rather than regaining ground lost. For example, the domestic auto industry had begun to concentrate on light trucks, ceding to European and Japanese producers important segments of the car market. The machine tool industry had abandoned significant parts of the standard machine market, reorienting efforts toward special machines and the design of their use by customers. Thus, whether or not they surpassed previous records, in sector after sector the increased openness of the U.S. goods market to products produced elsewhere had fundamentally changed the level and, just as important, the predictability of demand facing onshore U.S. producers.

The effect of this sea change is controversial. One view, which this writer finds optimistic to the point of Panglossism, holds that the 1970s ushered in the long-predicted internationalization of manufacturing, with routine operations besieged by competition from (and hence ceded to) poorer nations, with the richer nations compelled by a basically progressive logic to reposition their economies toward (as it is variously characterized) high-end, high value-added, information-intensive, and more flexible products and production methods.[1] Fordism was dead, but "flexible specialization" held promise of yet greater wealth, as technology enabled nimble, market-sensitive onshore producers to satisfy the increasingly finicky and variegated demands of richer and richer U.S. consumers.

What is not controversial is that the loss of predictable mass markets jolted the Fordist order. But the balance of the true story is far more sobering than the wishful thinking behind "progressive internationalization." Producing in shorter runs, many U.S. manufacturers suffered falling productivity or, at best, falling rates of productivity advance as expensive special-purpose machinery lay idle more of the time. More flexible equipment was available, but it was expensive and hard to cost-justify in the face of unpredictable demand. As a group, large firms took steps to retrench in their own operations back to more stable core businesses, outsourcing parts—and, they hoped, much of the demand instability—to their suppliers. At the same time, with profit rates low and falling, many implemented far-ranging cost-reduction strategies, often including branch plant siting aimed at securing lower-wage, nonunion labor. Indeed, while there has been frequent mention of, and more than a few books lamenting, the deindustrialization of America, the more pervasive trend, especially since 1979, is deunionization. As table 8.1 shows, while 1.2 million (6 percent) U.S. manufacturing jobs were lost in the 1980s, union job loss was more than 2.5 million (38 percent); nonunion manufacturing employment was *up* by 1.3 million (10 percent).

A Changed Manufacturing Mix

The combined result of all of this is extremely clear in the data, yet, apart from the much-noted productivity slowdown, little studied or reported. First, there was a huge change in the distribution of the manufacturing project as between large and small firms and, in large firms, between large and small plants. Between 1972 and 1992, plants with 500 or more employees shed almost three million employees, but plants with fewer than 500 added nearly two million (U.S. DOC 1972, 1992).

Second, forced to bear more and more of the instability of demand, the gap between smaller and larger plants' productivity—independent of the owning firms' size—began to grow. Value added per employee-hour was 20 percent lower in small plants (those with fewer than 500 workers) than in large plants in 1967; by 1977, it was 27 percent lower; and by 1987, more than

TABLE 8.1. Manufacturing Employment: Deindustrialization or Deunionization? (in millions of employees)

	Total	Union	Nonunion
1980	20.285	6.728	13.557
1990	19.064	4.197	14.867
Change	−1.221	−2.531	+1.310
Percent change	−6.0	−37.6	+9.7

Source: U.S. Bureau of Labor Statistics, *Employment and Earnings,* May 1982 and 1992.

31 percent lower. Not surprisingly, the small firm–large firm gap in workers' wages, as measured by payroll per hour worked, grew apace (fig. 8.1).

Indeed, the much-discussed productivity slowdown was, in significant part, the result of the change in the manufacturing plant mix. In plants with 500 or more workers, annual growth in value added per employee-hour slipped from 4.1 percent in the 1947 through 1967 period to 2.2 percent in the 1967 through 1987 period, a decline of 46 percent. But in sub-500-employee shops, the drop was from a 3.4 to a 1.6 percent annual rate, a decline of 53 percent (U.S. DOC 1947, 1967, 1977, 1987). With small plants accounting for half of manufacturing employment in 1967 but two-thirds in 1987, their more sharply declining rate of productivity advance added significantly to the overall slowdown.

Sources of Small Firm Performance Problems

The schedule instability that large firms imposed on many of their suppliers was not the only culprit in the latter's productivity problems. Just as they often went after unions as a way to control their own wage bills, large firms also squeezed their suppliers' profits, both by requiring regular price cuts over the life of multiyear orders and by forcing suppliers to share cost reductions resulting from lower-than-expected material costs and from productivity gains with them through lower piece prices. Suppliers, reasonably enough, reacted by breaking or avoiding unions, imposing pay cuts or freezes, and wherever possible avoiding new capital investments. By the late 1980s, as a result, smaller U.S. manufacturers were far behind both large U.S. firms and their offshore small-firm counterparts in their use of advanced manufacturing technologies (table 8.2)

The previously noted growth in the large firm–small firm productivity and pay gaps is probably one result. Another, less obvious one is the declining

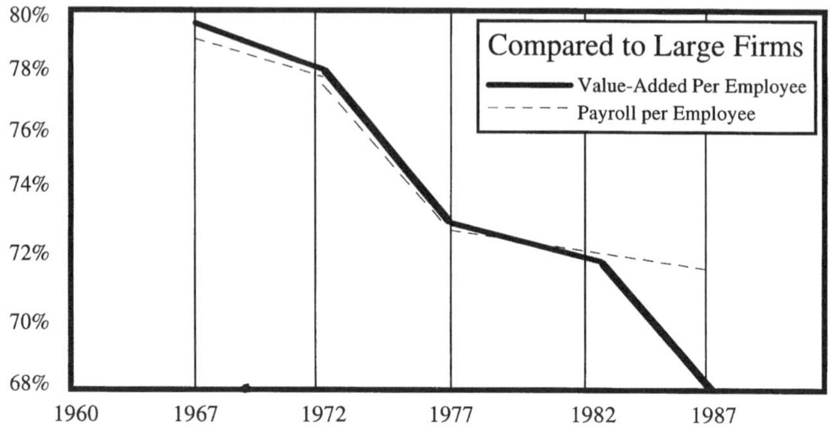

Fig. 8.1. Lower productivity holds down wages of manufacturing work force. (From U.S. Department of Congress, *Census of Manufactures,* 1967, 1972, 1982, 1987.)

technological homogeneity of different-sized firms in the value chain underlying manufactured products. By the 1980s, this began to affect not just the extent, but the nature, of large-firm outsourcing. In particular, more and more small firms were thought of less as lower-cost specialists in their products than as low-investment, low-overhead, build-to-blueprint shops. More and more of U.S. manufacturing, in a sense, was turned into a casualized service economy. While it may seem ironic amid all the clatter about partnership, longer-term contracting, and even emerging American *keiretsus*, this is what has occurred. And as a result, fewer and fewer small firms are in strong bargaining positions vis-à-vis their customers. Thus the squeeze on their profits and their employees' living standards continues.

TABLE 8.2. Percentage of Firms Using Technology, by Employment Size

	United States		Japan	
Technology/Employment Size	**50–499**	**500+**	**1–299**	**300+**
NC/CNC	39.6%	69.8%	57.4%	79.4%
CAD/CAE	36.3	82.6	39.1	75.2
Automated Inspection	10.5	44.3	30.1	66.7
Machining Cells	9.1	35.9	39.4	67.4
Robots				
Pick & place	5.5	43.3	22.6	62.2
Other	3.9	35.0	8.3	41.1

Source: U.S. 1987 data: U.S. Bureau of Census, *Manufacturing Technology 1988;* Japan 1988 data: MITI Small & Medium Enterprise Agency, *Small Business in Japan 1990.*

To some extent, the shift of manufacturing employment and value added from large to small firms is in evidence in nearly all advanced market economies. The loss of predictable demand and the advent of technologies that do not require large-scale operations have affected all market economies. Only in the United States, however, are the results so stark and perverse. In much of northern Europe, for example, wage disparities between large and small firms are—by law, contract, or custom—far more modest than in the United States. As a result, the inter-size group mix shift has little effect on workers' earnings. Moreover, the shift is less pronounced, since much less work is assigned to suppliers purely to take advantage of their lower labor costs. More of it is based on suppliers' competencies, and large firms are afraid to squeeze too hard on price lest their suppliers underinvest in those competencies. Less squeezed, European and Japanese suppliers have invested more than their American counterparts and, as a result, the widening large firm–small firm productivity gap afflicting the United States has not occurred.[2]

The Distribution of Small Firm Performance

Economists and lay business observers alike tend to think of market economies as ongoing experiments in social Darwinism. "Good" employees find jobs and get paid more than "bad" ones. Better firms grow; less good performers languish. Wages and profits send signals that attract labor-power and investment, and the cycle repeats, with the good guys winning. Indeed, much of the ideological appeal of markets is that the victory of the good over the less good means that, over time, the average improves as better firms take business away from less adept competitors and as less-good workers imitate the education and work-habit behaviors of their better-paid colleagues.

Implicit in all this is that good performance is rewarded. Firms that invest in new equipment, train their employees, manage their inventories, and deliver on time to their customers, to pick a few attributes, succeed where many of their less-good counterparts do not. It therefore comes as a big surprise to most economists that primary data on the performance of manufacturing firms show no such clear pattern. This writer is part of a group at the Ann Arbor, Michigan, Industrial Technology Institute that maintains a data base, updated annually since early 1992, on more than 1,000 small U.S. manufacturers in both high-volume, mainly build-to-blueprint industries such as metal stamping and plastic products as well as low-volume, generally engineer-to-order industries such as tooling and special machines.[3] Work to date convinces us of a much less savory reality. Our data on manufacturing performance show that firms making identical products often employ radi-

TABLE 8.3. Performance of High-Volume Metalforming Shops, 1993

	Lowest Value		
Measure	**Top 10%**	**Top 25%**	**Top 75%**
Value-added per employee	$122,400	$91,100	$55,000
Total inventory turns	34	21	8
On-time delivery	99.3%	97.5%	88.0%
Training per employee	$400	$147	$17

Source: Performance Benchmarking database.

ally different production recipes, as table 8.3—reporting some 1993 results for 60 high-volume metalforming shops—makes clear.

Even among similar firms facing similar customer demands, there are huge disparities in performance. Some metalformers turn over their inventory every few weeks, others at most every other month. Some make fewer than 1 percent of their deliveries late, while others are late more than one delivery in eight. Some spend more than 2 percent of their payroll, year in and year out, on training their workers; others, even in the same industry and local labor market, never spend anything. Finally, productivity (as measured by value added per employee) ranges from as little as $40,000 to as much as $140,000 (and from as little as 130 percent to as much as 300 percent of payroll).

Differential performance on the measures shown in table 8.3, and on more than 100 others on which our group rates data-base members, follows certain broad outlines. While very few firms are good (score in the top quartile) across all or even most of the measures we calculate, high performance across measures clusters in recognizable ways. Recipes range from a low-road or "sweating" approach of low wages, skills, capital spending, and automation to a high-road strategy of high wages, skills, investment, and technology. Low-road shops may be thought of, essentially, as renting machine and machine-operator time to their customers. High-road shops include those that make products that can only be made by skilled labor (e.g., special tools, dies, jigs, and fixtures) as well as a few that make more routine products but that through accidents of history (e.g., unionization or location in a high-wage labor market) chose the recipe even though otherwise-identical competitors did not. Finally, about one firm in three is using a hybrid recipe that we characterize as "lean commodity" manufacturing. Like the low-road shops, they are unable to represent themselves to their customers as sufficiently distinctive in product or process to command premium product prices. Unlike their low road counterparts, however, they are more self-conscious and improvement-oriented, often using Japanese-style methods (just-in-time, worker self-inspection, problem-solving groups, etc.) to root out waste and reduce costs. On wages, training per employee, and value added per employee, lean commodity firms score between low- and high-road shops, though on average substantially closer to the low.

A Theory of Manufacturing Performance

We have also begun to integrate this inferred taxonomy of high-road, low-road, and lean commodity firms with a theory of manufacturing improvement, based on analysis of the same database.

The theory is that firms can be ranked on three scales. The first scale, the "good", endpoint of which we call *systematic*, captures the large set of activities that distinguish shops that are haphazard from those that are deliberate—tracking potentially dozens of indicators (scrap, dimensional conformance, setup time, schedule integrity, machine uptime, causes of equipment breakdown and tool wear, etc.) and engaging in a wide variety of corrective behaviors (total quality management, problem-solving groups, work teams, statistical quality assurance, just-in-time, pay-for-performance, documented preventive maintenance, etc.) Lean commodity firms are far more systematic—and far more focused on getting even more so—than low-road shops.[4]

The second scale, on which lean commodity firms typically rank only a little higher than low-road firms, we call *modernity*. High scores on this scale require not just hardware and software that automate the business, scheduling, and quality functions of systematic firms (e.g., manufacturing resources planning software, bar coding, electronic data interchange, statistical process control) but product-development and shop-floor process automation (e.g., depending on industry, computer-controlled machines, computer-aided design workstations and software, programmable machine controllers, etc.). Low-road firms never, and lean commodity firms seldom, score well on this scale. High-road firms tend to score high on this scale, reflecting (we believe) the incentive their higher wages give them both to automate and to train the remaining work force to use the automation to good effect. There is, perhaps surprisingly, not a great deal of correlation between high scores on the systematic and modern scales. Modern shops are often quite chaotic, and (as already noted) most systematic shops are not particularly modern.[5] The third, most problematic, and most interesting scale we call *distinctiveness*. In light of our earlier story on the antitechnological and wage-cutting effects of large firms' price-squeezing their suppliers, it's good to be distinctive, since we define it as having attributes that permit a firm to charge a premium for what it sells. The premium-commanding distinctiveness might be in products (new, proprietary, or design-intensive), processes (unique, able to achieve ultra-tight tolerances or handle exotic materials), or craftsmanship. Our earlier remarks about European and Japanese suppliers vis-à-vis U.S. suppliers might be supplemented by the strong hypothesis that more of the former are distinctive, with that distinctiveness conferring bargaining power, profitability, and the ability to sustain high wages in a virtuous circle. Moreover, their distinctiveness carries an additional, infinitely precious reward for their customers: freed from the need to carry large overheads in the design and engineering of their suppliers' parts, large firms can focus on charting end-use markets and on organizing their own and their suppliers' efforts on behalf of rapid new product introductions to satisfy those markets. By being first with those products, large firms can often charge premium prices, netting extra profits that further obviate the need to squeeze their suppliers. (Sadly, these observations have probably been rendered less and less accurate during Europe's tilt away from social democracy.)

Relatively few firms score highly on our distinctiveness scale, reflecting, we believe, the fact that relatively few small U.S. shops are distinctive, particularly when capital goods (about 15 percent of the small-firm manufacturing economy) are excluded from the analysis. One would think that scoring high on distinctiveness would be associated with large advantages in markups and hence profitability. In fact, however, while being modern is associated with high productivity, and high productivity with high profits, high scores on the distinctiveness scale are not associated, in either direction, with high profits. We have two hypotheses in this regard. First, we believe that few large U.S. firms want distinctive suppliers. For all the talk of concurrent engineering, system sourcing, and other indicators of value-chain marriage, most customers want one-night stands so that they can keep the lion's share of the value chain's surplus for themselves. Second, we think that many distinctive firms are suffering from eroding distinctiveness. Much of their past advantage was based on the craft element of the distinctive scale, and an important attribute of programmable technology has been to transform many complex machining and assembly operations into science rather than black art. While skilled hands still matter, increasingly they are not enough by themselves to make firms stand out among their peers. According to this hypothesis, many high-wage, high-craft firms are being beaten out by lower-wage, systematic, and modern firms.

Finally, there is no correlation between scores on any of our three scales and sales growth. It is this that is, perhaps, most damning for the social Darwinist view. If one rules out the implausible possibility that "bad" firms only seem to be doing passably well for now but are about to stop growing and

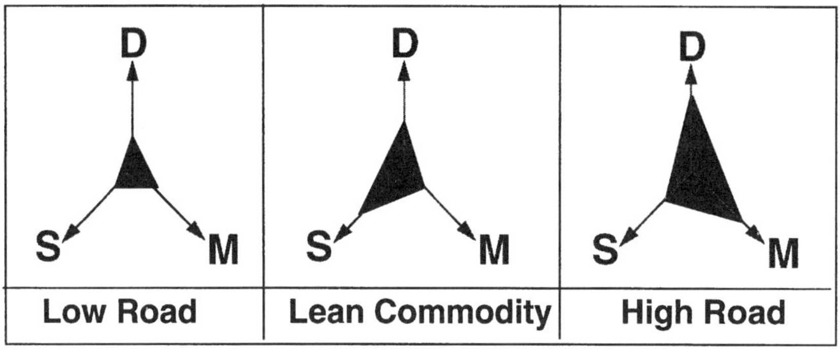

Fig. 8.2. Performance scales and manufacturing strategies

walk to their doom off the proverbial cliff, then one must conclude that markets do not reliably drive low-road manufacturers out of business but, instead, often reward their sweating strategy as much, or nearly as much, as strategies undertaken by "good" firms.

Implications for Policy

If in most industries firms can choose from among recipes with very different mixes of wages, skill, technology, training, and indeed business and manufacturing discipline without incurring significant growth or profitability penalties, there is little basis for expecting that markets alone will call forth high-performance behaviors. On the other hand, the low frequency of such behaviors may be bad for the U.S. economy, resulting in lower wages, higher large-firm engineering overheads, less technical change, and a composition of output that has too few products that fare well in international trade.

It is this possibility that provides the economic justification for policy intervention to shape the incentives that influence manufacturers' behavior. Based on our earlier argument, we favor policies that aim directly at migrating the distribution of smaller manufacturers toward the high-road tail. After all, as we have argued, it is the lag in small shops' technology, productivity, and wages that explains much, if not most, of the overall U.S. productivity growth problem, living standard stagnation, and trade deficit in manufactures.

What are the elements of a policy that aims to change the distribution of smaller manufacturers as between low-road, lean commodity, and high-road recipes? In keeping with the metaphor of roads, it may be useful to classify these elements as roadways (positive initiatives, incentives, and indicative signals) and roadblocks (disincentives and punitive regulation).

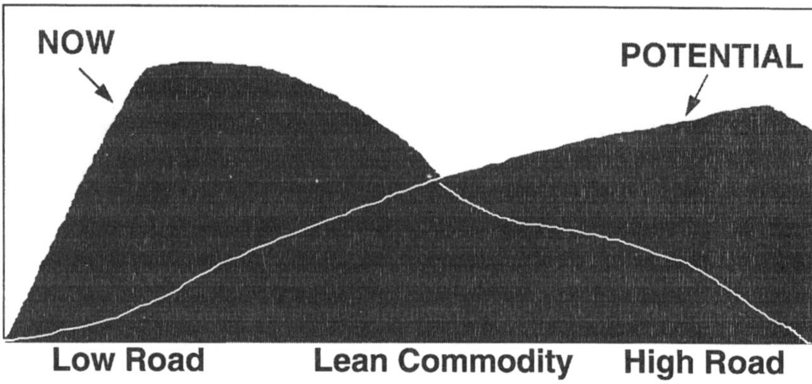

Fig. 8.3. Distribution of small manufacturer performance

Roadways to High Performance

First and foremost, robust economic growth is required. Slow growth heightens price competition, which in turn reduces the resources available to implement high-road behaviors, many of which (automation, new product launches, etc.) have high front-end costs.

Second, most of the increased demand must be reserved for onshore producers. Otherwise, much of the impact of economic recovery will leak out; if this happens, it will not lead to reduced price-squeezing of U.S. suppliers. This is probably best pursued by conditioning access to U.S. goods markets on reduced bilateral deficits rather than on industry-by-industry restraint agreements. The latter do make sense, however, for industries in which the supplier base has many firms that are on the fence about which recipe to choose: auto and appliances, among others, meet this test; aerospace and machine tools probably do not.

Third, policy should aim to increase the supply of workers skilled in tasks required by high-road small shops. Rather than respond to average or general demands, the content of vocational training should be driven by the needs of only the highest-performing shops. For example, because product design is often a key differentiator of high-road firms, there should be publicly supported design apprenticeship grants available to small firms. Training should also equip workers in low-road and lean commodity shops to migrate to better-paying jobs in high-road firms. Such training should be financed by a tax that is borne most by firms that train least. This could mean a flat payroll tax paid only on, say, the first $10,000 of annual earnings per employee or, as the Clinton administration considered in its comparatively heady first year, by imposing a tax of 1.5 percent of payroll, which firms may escape if they are already spending that much or more on qualifying shop employee training.

Fourth, investment in modern machinery and controls should be encouraged. One approach would be an investment tax credit applied only to computer-controlled design and manufacturing equipment. Since high-road shops are far more capital-intensive than others, such a credit would reduce their factor costs disproportionately. Software costs should be excluded to deny the subsidy to lean commodity shops with automated business, but not shop floor, processes. There should be no investment tax credit on structures, since greenfield plants are often part of the maturity-avoidant strategies of low-road and lean commodity firms. An especially generous credit should be considered for qualifying investments in computer numerically controlled (CNC) mills and lathes, the key equipment used in the production of tools, dies, molds, jigs, and fixtures. This is in recognition of the intrinsically high-road nature of such production and of the role that more efficient production of these capital goods plays in reducing the product cycles of major traded-goods industries.

Fifth, health reform should be shaped to remove the huge cost penalties faced by mature firms, among which high-road behaviors are most prevalent. Firms should not pay higher premiums because of the age of their active or the number of their retired workers. In addition, to the extent that a reformed insurance system is financed by payroll taxes on employers, such a tax should be based not on all of payroll (which is higher per employee in high-road firms), but on the product of annual employee-hours and the legal minimum wage.

Sixth and last, there needs to be a much-expanded system of industrial extension programs that help small manufacturers upgrade their operations. But to serve the goal of high performance, the extension system should not be a universal entitlement for small firms, but instead be carefully targeted to assist high-road shops and those committed to joining them on that road. The key is to site these programs in places that are host to large numbers of high-road suppliers to important, traded-goods industries. Not only would doing so constitute a larger subsidy to high-road firms, it would also upgrade manufacturing as a whole more efficiently, since technology and technical know-how diffuse faster in densely agglomerated regions that host both suppliers and their major large-firm customers.[6] Finally, the explicit substance of extension programs' assistance should be to assist qualifying firms' efforts to prosper using high-wage, high-skill, high-technology recipes and to mobilize their resources not just to offer low prices to their customers but, more important, to become more distinctive and hence less exposed to price-squeezing. This has obvious implications for the role large firms should play in the design and operation of such programs.

Roadblocks to Low Performance

First, the low road should be partially blocked by labor law reforms that make union-busting harder and that speed the implementation of first contracts at plants where unions have won recognition elections. Unions raise wages, and higher wages are an essential component of high-road recipes as well as important to maintaining domestic demand. To ensure that unions deliver for their members rather than merely collect dues, automatic deduction of dues from workers' paychecks (checkoff) should be barred.

Second, there should be a much higher minimum wage. In order to speed the economy's adjustment to a distribution with many more high-road firms, the new minimum should be announced as taking effect at a time certain several years out. Just as EC 1992 galvanized European companies to take action in advance of a known deadline, telling American firms that the minimum hourly cost of labor will be, say, "$9.09 in '99" would spark a wave of recipe-changing. One aspect of that wave will be a boom in labor saving capital goods, which (fortuitously) are produced at wages far above the average. Such capital goods, many oriented toward service industries, might also be expected to enjoy huge export markets.

Third, the training policies discussed earlier in effect punish low-road firms, the vast majority of which do little or no worker training.

Fourth, federal legislation is needed to end the prerogative of state and local governments to abate property taxes on new manufacturing facilities in industries that have operated for more than, say, five years at less than 85 percent of capacity. The vast majority of new plants are being built in rural and exurban areas by low-road and lean commodity firms, or by firms replacing high-road plants with low-road or lean commodity branch plants. Making new capacity more, rather than less, expensive than mature capacity would thus discourage low-road behavior.

Conclusion: Regional Policy as High Performance Industrial Policy?

Policies such as these are sure to have uneven inter- and intraregional impacts. Mature industrial areas—including many upper Midwest cities—host a disproportionate share of high-road firms, while the Southeast and most rural areas exhibit largely low-road and lean commodity manufacturing populations. While this may give rise to political objections, much as the regionally unequal distribution of federal prime defense contracts since World War II has done, such an approach also has at least three notable virtues. First, it promises major benefits in terms of international competitiveness. The Midwest produces most of the nation's manufactured exports, and its industries' loss of domestic market share is the primary culprit in higher manufactured imports since the late 1970s.

Second, an industrial policy focused on giving incentives for high performance in smaller manufacturers would constitute an active urban, or at least metropolitan, strategy as well. Bicoastal defense contracting and the encouragement of greenfield manufacturing investment have subsidized rural areas and distant suburbs at the expense of cities and their inner-ring suburbs. Cities remain home to millions of manufacturing jobs; many companies remain dependent on the rich pool of skills only available in mature metropolitan labor markets. As we've noted, metropolitan manufacturing is higher-wage and more capital-, technology-, and skills-intensive. But since the 1970s, metropolitan manufacturing has been strongly challenged by nonmetropolitan low-road and lean commodity firms. From 1978 to 1988 (peak to peak, for manufacturing), the United States lost 1.93 million (9.5 percent) of its manufacturing jobs. Just 17 cities—Chicago, Gary, Detroit, Flint, Newark, Buffalo, New York City, Cleveland, Toledo, Youngstown, Allentown, Bethlehem, Philadelphia, Pittsburgh, Providence, Houston, and Milwaukee—lost 1.325 million, 70 percent of that total, even though in 1978 they hosted just 20 percent of U.S. manufacturing jobs (table 8.4).

Last, such a strategy has a good chance of splitting business opinion along politically promising lines. Because today business usually lobbies with one voice, its positions are often least common-denominator endorsements of anything that holds down labor costs. The problem is that such an approach is actually a subsidy to low-road firms and hence a relative burden to high-road companies. A policy that combines support for smaller firms and initiatives that provide extra help for those with the highest labor costs might help coalesce a fraction of capital in support of a high-road social contract. It

TABLE 8.4. Millions of Manufacturing Employees

	1978	1988	Change	%Change
Seventeen cities	4.091	2.766	−1.325	−32.4%
Everywhere else	16.228	15.623	−.605	−3.7%
Total	20.319	18.389	−1.930	−9.5%

Source: U.S. Bureau of Labor Statistics, *Employment and Earnings,* May 1980 and 1990 issues.

is the absence of such a fraction today that allows the pursuit of wage-led economic growth to appear a parochial trade union demand, rather than the general interest it could and should be.

NOTES

1. See, among others, Reich 1991.

2. See, among others, Amin 1989, 24 and Rey 1991. On Japan, see MITI 1985; Bruch and Hiemenz 1984, esp. 26.

3. The Performance Benchmarking database is supported by the U.S. Department of Commerce's National Institute of Standards and Technology (NIST) and the Michigan Strategic Fund. Data are collected annually from 11 industry panels using detailed, heavily piloted 200-question protocols, with extensive postquestionnaire checkbacks with participants to ensure data quality and accuracy.

4. Our analysis, led by Edith Wiarda, uses a numerical scale on which firms can score up to 56 points if they have sophisticated cost estimating; repair or replace tooling on a scheduled and preventive basis; train widely in and quickly correct errors detected by statistical quality assurance; and routinely track data on quoting accuracy, lots and deliveries on schedule, the proportion of machine hours actually running parts, and a variety of measures of scrap, rejects, and inspection effort. Firms scored from 6.6 to 46.0 points on this scale. Each additional 10 points is associated with 5-percentage-point reductions in both lots that have to be expedited and in late deliveries, and with 1-point drops in both scrap and customer reject rates. In short, systematic behaviors get firms schedule integrity, delivery, and quality gains.

5. Our numerical scale on modernity gives points for a wide variety of hardware and software, as well as for newer and more automated machinery. Interestingly, one of the most powerful variables in the scale is the proportion of employees using a computer or programmable machine controller at least weekly, which is more a measure of how widely, rather than deeply, programmable technology is used. Each additional 10 points on employee computer/controller use is associated with about $2,700 more in value added per employee. While low road firms that score poorly on this scale sometimes make high profits through sweating, every $10,000 increase in value added per employee is associated with nearly 2 percentage points more profit/sales.

6. Such a siting strategy could be driven by determining the set of large, traded-goods sectors supporting large numbers of smaller suppliers. Once known, the subset whose suppliers are most densely located and intraregionally linked with their customers can be identified. The winners turn out to be 22 metropolitan areas. Even after correcting for industry mix, earnings per manufacturing employee in these 22 places average 20 percent higher than in the rest of the manufacturing. Every one of the 22 is in a state that boasts a disproportionate number of small tooling and industrial pattern shops (SICs 3543–44–45; hereafter, tooling). While tooling has just 200,000 employees nationwide, its special status comes from the fact that because patterns, dies, molds, jigs, fixtures, and custom gauges define products, they also gate how cheaply, quickly, and accurately products can be manufactured. Luckily, modernizing tooling involves a well-defined recipe: CNC mills removing metal by following cutter paths generated by CAD/

CAM software from CAD product files. Every extension center should be in the business of CAD-CAM/CNC in tooling; the only variance should be in the size of the field staff engaged in it. The pro–high-road extension siting argument is developed in detail in Luria et al. 1994.

REFERENCES

Amin, Ash. 1989. Flexible Specialization and Small Firms in Italy: Myths and Realities. *Antipode* 21, no. 1: 13–34

Bruch, M., and U. Hiemenz. 1984. *Small- and Medium-Scale Industries in ASEAN Countries: Agents or Victims of Economic Development.* London: Westview.

Luria, D., R. J. Cole, A. Baum, with E. Wiarda, C. Treado, and M. Grueber. 1994. Fixing the Manufacturing Base: The Allocation of Manufacturing Extension. *Journal of Policy Analysis and Management* 13, no 3: 571–80.

Ministry of International Trade and Industry (MITI). 1985. White Paper on Small and Medium Enterprises in Japan 1985. Background Paper #B1-58, Tokyo.

Reich, Robert. 1991. *The Work of Nations.* New York: Knopf.

Rey, G. 1991. Small Firms: A Profile of Their Evolution, 1981–85. In *Small Firms and Industrial Districts in Italy,* ed. E. Goodman. London: Routledge.

U.S. Bureau of Labor Statistics. Various issues. *Employment and Earnings.*

U.S. Department of Commerce, Bureau of the Census. 1973, 1978, 1979. *Annual Survey of Manufactures.* Washington, D.C.: Government Printing Office.

U.S. Department of Commerce, Bureau of the Census. 1947, 1967, 1972, 1977, 1987. *Census of Manufactures.* Washington, D.C.: Government Printing Office.

U.S. Department of Commerce, Bureau of the Census. 1992. *County Business Patterns—United States.* Washington, D.C.: Government Printing Office.

CHAPTER 9

U.S. Competitiveness and Economic Growth

Candace Howes

In the introduction to this book, Howes and Singh (1999) argued that there are good analytical and empirical reasons for the view that relative productivity growth and the relative competitiveness of a country's export sector matter profoundly to its overall economic performance. With so much trade based on nonprice competitiveness, the trade balance can rarely be achieved solely through exchange rate manipulation or only at great cost in terms of employment and real income growth. Moreover, greater productivity growth abroad, as a result of faster technical progress there, is likely to have negative consequences for productivity growth in the home economy unless corrective measures are taken to enhance the country's technological capabilities. Thus even an advanced country cannot afford to ignore its international competitive position if it wishes to improve its standard of living in the long run.

Contrary to Krugman's (1994) argument, competitiveness matters. Otherwise, in a world of wage-price inflexibility, external balance is achieved only through adjustments in a country's rate of growth relative to that of its trading partners. Furthermore, due to cumulative causation, a country that is investing and innovating at a slower rate than its competitors will fall even further behind, requiring slower and slower relative growth. Real incomes will fall, inequality will increase, and employment will stagnate.

In this chapter, I argue that the United States has such a relative competitiveness problem, which, because of wage-price inflexibility and the inadequacy of exchange rate adjustment, is manifest in a structural decline in its current account balance. Ultimately, it cannot sustain its current account balance without slowing growth.

Second, I provide evidence that the maximum sustainable rate of growth has declined over time, due both to a structural decline in the current account and to the relative slowing of the growth of rest-of-world demand. Trends in the income elasticities of demand provide evidence of declining relative competitiveness of U.S. goods, especially nonprice competitiveness.

The structural decline in the balance of payments—manifest in increasingly unfavorable income elasticities of demand for imports and exports—is due specifically to:

An uncompetitive manufacturing sector;

A service sector that, despite its dynamism, cannot make up for the decline in manufacturing;

A structural decline in net capital income flows, due largely to the financial liberalization of world capital markets and the uncompetitiveness of the manufacturing sector.

Finally I argue that the balance of payments problem cannot be remedied by a currency depreciation because while the United States does not face wage-price rigidity quite to the degree that Europe does, it is nonetheless incapable, due to the deregulation of international capital markets, of depreciating its currency. If anything, pressures will become stronger for appreciation. While other elements of relative unit costs such as wages might improve the relative price position of U.S. exports, the United States is unable to increase productivity growth, in part due to its slow rate of growth. But even with regard to U.S. wages there is evidence that they are already stagnating at what is possibly the politically acceptable limit.

In the following section I briefly discuss why declining relative competitiveness—manifested by unfavorable trends in income elasticities—acts as a constraint on growth. I then discuss how relative price adjustments could in theory remedy the problem, but in practice probably will not. Next I analyze trends in the U.S. current account since 1960 to illustrate the central argument that U.S. growth is constrained by structural deterioration in the current account that is due to an uncompetitive manufacturing sector. I then decompose trends in components of the current account to show that there is no magic bullet, such as service exports, that will offset negative trends in manufacturing trade. In the last section I provide a simple simulation of probable trends in the current account balance and the effect they will have on growth.

Why uncompetitive exports cause a current account deficit

The United States cannot for long maintain a sustainable current account balance without either slowing growth or depreciating its currency or both.[1] Unfortunately, whereas in the past relative price adjustments have partially accommodated the competitiveness problem, they may no longer be able to play that role.

The structural deterioration in the sustainable current account has arisen due to a number of factors. First, the United States has become less competitive relative to its trading partners in the production of goods that compete on the basis of nonprice factors. As a consequence, the income elasticity of demand for U.S. imports exceeds that for U.S. exports, and there is some evidence that the difference is getting larger over time. Second, the rate of growth of demand of U.S. trading partners has slowed relative to that of the United States. Third, while there is some evidence that demand responsiveness to trends in relative prices may have increased, the United States seems virtually powerless to depreciate its currency.

Since Houthakker and Magee (1969) found for the 1950s and 1960s that the income elasticity of demand for U.S. imports exceeded that of exports, it has generally been understood that, *for similar rates of growth of foreign and U.S. demand,* U.S. imports will grow faster than exports. This will be true unless foreign demand for U.S. goods can be sustained through a fall in the relative price of U.S. goods. The rate at which the relative price must fall depends not only on the magnitude of the difference between the income elasticities, but also on the degree to which demand is responsive to prices, that is, the price elasticities of demand for U.S. imports and U.S. exports.

A simple arithmetic example illustrates this point. Suppose the income elasticity of demand for U.S. imports is 1.5, while that for exports is 1.0, and the price elasticity for both imports and exports is −1.0. Then, other things being equal, as long as the demand for U.S. exports grows 50 percent faster than does U.S. demand for imports, there will be no trend deterioration in the trade balance. If, however, the rest of the world grows at the same rate as the United States, and if demand grows at the same rate as output, balanced trade can be maintained only by a steady fall in relative prices. If demand is growing worldwide at the rate of 3 percent a year, relative U.S. prices would have to fall by 0.75 percent annually. A 4 percent U.S. growth rate would require an annual decline in relative prices of 1 percent. The less responsive demand is to relative prices, the more prices will have to fall, annually.

Fortunately, for several decades the United States was able to ignore the implications of unfavorable income elasticities because the rate of growth of world demand greatly exceeded U.S. demand growth and because U.S. prices (as measured by the real effective exchange rate) were growing more slowly than the rest of the world. Thus the United States was able to maintain balance on its current account for three decades after the end of the World War II.

Given the right conditions, it is possible for unfavorable income elasticities to improve over time.[2] It is therefore useful to consider what determines income elasticities and whether the conditions for improvement exist in the United States. Kaldor (1981) argued that income elasticities were a measure of the competitiveness of a country's tradable goods along nonprice dimensions. Thus, as the product quality and rate of innovation rise relative to other countries, the income elasticity of demand for exports should rise and for imports fall.

As income elasticities are aggregate concepts, reflecting the average response to the quality of a country's particular mix of exports, it is also true that income elasticities can change as the *mix* of exports changes. For example, as a country's productive capabilities shift to a range of products that compete more on nonprice terms (as will be the case for countries with high rates of investment), trends in income rather than price will be more important in determining the growth of net exports. Thus, if the relative quality and innovativeness of a country's products are rising and if the share of nonprice competing goods in its total exports is rising, while falling in its import mix, then the ratio of a country's import to export elasticities should improve (that is, fall).[3]

As I will show in the section "Will Services Offset the Manufacturing Deficit," there is good reason to expect (as well as empirical evidence to support the hypothesis) that the income elasticities of demand for U.S. products are becoming less favorable over time, reflecting a decline in relative export competitiveness.

Although income elasticities have become relatively more important over time in determining rates of growth of net exports, there is evidence that relative prices have historically played a significant role, at least in short-term fluctuations in the U.S. current account. It is perhaps the case that since the United States has had greater wage-price flexibility than was the case in much of Europe, price-competing products held a greater share in both its export and import mix.

There are several mechanisms through which the relative price of U.S. goods can fall. First, real currency depreciation, assuming the depreciation is passed through to lower prices for exports in the foreign currency and higher prices for imports in the domestic market, will, other things being equal, lower the relative price of U.S. exports. Second, if other things such as the exchange rate and relative productivity are held constant, declining relative compensation and relative input prices will lower relative export prices. Conversely, increasing relative productivity, holding relative compensation and other input prices constant, will have a similar effect. Of course, some combination of all three is most likely, but currency depreciation requires the least in terms of institutional change. The trend in relative real unit labor costs summarizes the combined effects of trends in the exchange rate, relative compensation, and relative productivity growth.[4]

As the next section shows, trends in relative prices (as measured by real effective exchange rates based on relative unit labor costs) during the 1980s caused large movements in the current account balance. However, given the apparently intractable problem of slow productivity growth, the politically unacceptable stagnation in real wages, and the inability of the United States to devalue its currency, relative price adjustments are now less likely to remedy the external imbalance.

U.S. External Performance and Relative Competitiveness

Turning to trends in U.S. external performance, we find that there is considerable evidence for the proposition that U.S. competitiveness in nonprice competing products has eroded. Further, there is evidence that U.S. exports tend to compete more on the basis of price than in the past. However, the United States is limited in its capacity to improve relative price performance through exchange rate manipulation. These circumstances now make it virtually impossible for the United States to maintain external balance at a reasonable rate of growth. Table 9.1 helps illustrate these points.

In the 1960s, the United States grew at a fairly rapid pace, just slightly below that of the rest of the world. At this rate of growth, the United States was able to maintain full employment and external balance simultaneously without resorting to currency depreciation. During this period, on average, U.S. compensation was growing at a slower pace than that of the rest of the world, and this was sufficient to more than offset the effect of its relatively slower productivity growth on unit labor costs. U.S. relative unit labor costs fell at an annual rate of about 5 percent in the first half of the decade and rose at a rate of about 3 percent for the second half, averaging about −1.7 percent a year over the entire period.[5]

The current account surplus had shrunk to 0.04 percent of GDP by 1969 (down from 0.5 percent in 1960 and 1 percent of GDP in 1964), and in 1970, the United States floated the dollar. The nominal effective value of the dollar fell by 6 percent a year between 1969 and 1973 (Hooper 1992).[6] Since relative U.S. compensation continued to fall in national currencies, and since U.S. productivity growth recovered somewhat, relative unit labor costs in dollars fell at an annual rate of approximately 10 percent (fig. 9.1). By 1973, then, relative US unit labor costs in dollars were at about the level they would have

TABLE 9.1. Indicators of Trends in U.S. Competitiveness, 1960–97 (in percentages)

	Current Account/GDP	Real U.S. Demand Growth	Growth of World Demand	U.S./World Growth	REER (RULC)	Unemployment Rate	Inflation Rate
1960–69	0.5	4.6	4.9	0.94	−1.7	4.8	2.5
1969–73	0.0	3.6	5.4	0.67	−10.0	5.4	5.4
1973–79	0.0	2.8	4.6	0.62	−3.6	6.8	8.1
1979–90	−1.8	2.4	3.7	0.64	−0.9	7.1	5.0
1990–97	−1.4	2.4	2.6	0.91	0.2	6.2	2.6

Source: Compiled by author from OECD Economic Outlook #63 Database diskette.

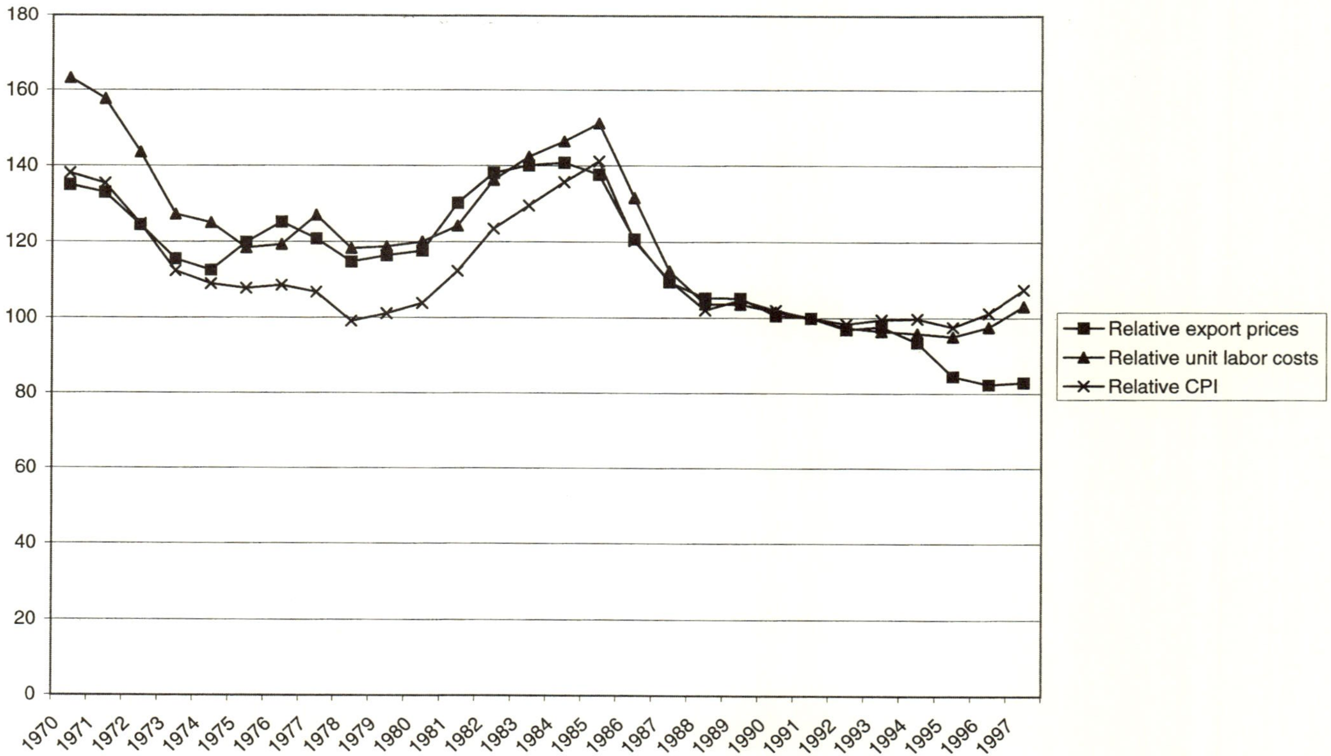

Fig. 9.1. Real effective exchange rates, 1970–97 (1990 = 100)

been if the trend decline of the early 1960s had continued through 1973. U.S. relative unit labor costs were 62 percent of their 1960 level, an average annual decline of about 3.6 percent for the 13-year period.

Looking at the period 1973 through 1979, we find that although relative compensation in national currency was falling at about the same rate as relative productivity, so that relative unit labor costs in national currency were falling only at a rate of about 1 percent a year, the exchange rate was depreciating at a rate sufficient to maintain the trend decline of 3.6 percent annually in relative unit labor costs right up to 1979.

Thus for the period 1960 to 1979, the following picture emerges. The United States was able to maintain external balance on average throughout the period under the following conditions: (1) relative unit labor costs were falling by almost 4 percent a year, (2) the rate of growth of U.S. demand was declining, both in absolute terms and relative to its trading partners. Over the period, the rate of growth of U.S. output fell from 4.6 percent in the 1960s to 2.8 percent in the later 1970s and from a rate that was 92 percent to a rate that was just 62 percent of the rate of growth of rest-of-world demand. The fact that current account balance required a steady decline in the relative and absolute rate of growth of U.S. demand over the period suggests that the income elasticity of demand for U.S. exports relative to imports must have been falling as well. In terms of nonprice competition, this would be evidence that U.S. competitiveness was falling steadily over the period.

With the growth rate of demand in the rest of the world slowing at the same time that U.S. competitiveness was eroding, the maximum rate of growth of demand that the United States was able to sustain, consistent with balance on the current account, fell 40 percent, from an annual rate of 4.6 percent in the early 1960s to 2.8 percent in the second half of the 1970s. Slower growth raised average unemployment rates 40 percent from 4.8 in the 1960s to an average of 6.8 percent for the late 1970s.

The 1980s marked a new era in current account performance for the United States. Despite what in retrospect appears to be a trend deterioration in competitiveness for the United States during the 1970s, whether or not the United States suffered from a competitiveness problem was not really a public concern until the 1980s. As is well known, during the first half of the 1980s, the U.S. external balance deteriorated so rapidly that by 1987 the current account deficit had reached 3.6 percent of GDP. Most economists agreed that the deterioration in the current account could be largely explained by the rapid appreciation of the dollar between 1980 and 1985, combined with the fact that U.S. demand was growing faster than that of the rest of the world (Helkie and Hooper 1988; Mann 1999). However, the fact that the current account was still in substantial deficit in 1988, despite the dollar having fallen back to its real value of 1980, sparked a huge debate about the causes of the deficit, especially the role that might have been played by declining U.S. relative competitiveness.

Several theories were offered to explain why the U.S. current account did not return to balance with the depreciation of the dollar. Some argued at the time that mismeasurement hid the fact that the dollar had not actually returned to its 1980 level (Cline 1989). Others suggested that it was simply a problem of lagged adjustment (Lawrence 1990). Still others argued that for the current account to be balanced the dollar needed to fall to the level it would have reached had it continued on the trend decline of the 1960s and 1970s (Blecker 1992, 1999). Finally, there was a question of whether the depreciation of the dollar was actually being passed through to import prices.[7]

Whatever the cause, the fact is that by 1990, two to three years after the dollar returned to its 1980 level, the current account deficit was still 1.6 percent of GDP. On average over the entire period from 1979 to 1990, the current account deficit had run 1.8 percent of GDP, despite the fact that U.S. growth lagged the rest of the world by about 40 percent.

Thus the 1980s differed from the 1970s in that although the United States was still growing slowly relative to the rest of the world, there was no trend decline in relative prices, such as had been characteristic of the 1960s and 1970s. In the absence of falling relative prices, the United States was no longer growing slowly enough to balance the external account. Even so, since the average rate of growth of its trading partners had trended down from the 1960s and 1970s, for the United States to grow relatively more slowly meant that its growth rate compared to the 1960s was cut almost by half. Consequently, average unemployment rates were 50 percent higher than in the 1960s.

Since 1990, the external balance has improved slightly, on average. The average is distorted by the effect of an exceptional transfer of $43 billion dollars in 1991 in payment for U.S. engagement in the Gulf War. But it is also true that since 1993, the U.S. current account deficit has not risen above 2 percent of GDP, despite the fact that the 3 percent annual growth of U.S. demand has been roughly equal to the rest of the world and despite real exchange rates actually beginning to appreciate.[8]

The main point to draw, however, from the trends since 1990, is that even if the United States grows at the same slow rate the rest of the world is now experiencing, it sustains a chronic current account deficit of at least 2 percent of GDP, as it has for the last five years. The expected appreciation of the dollar with the financial crisis in emerging markets should only exacerbate the problem. This is not to deny that there is some apparent change in the underlying trends such that while the United States had to grow more slowly than the rest of the world in the 1980s in order to keep its current account deficit below 2 percent of GDP, since 1990, it has been able to grow at about the same rate as the rest of the world without any real deterioration in the external balance. But the crucial point is that the rest of the world is now growing very slowly relative to historic trends, which constrains the United States to similarly slow growth, given its propensity to import more than it exports.

Will Service Exports Offset the Manufacturing Deficit?

Good net export performance on the part of services—a sector that had not been a significant component of the trade balance in the past—has raised the possibility that manufacturing competitiveness is no longer essential to achieve sustainable external balances. In this section, I decompose trends in the components of the current account in order to assess that possibility. I also consider the implications of the decline in net income flows for long-term trends in the external balance.

The first striking point about the U.S. current account balance is that it was virtually balanced until 1976 (figs. 9.2, 9.3). Between 1975 and 1980, the trade balance moved into deficit largely due to the increase in oil prices, but was offset by capital income flows sufficient to keep the current account roughly balanced. Since 1980, most of the components of the current account have deteriorated or at least not improved, with the exception of services. The balance in goods trade, which was said to have deteriorated in the early 1980s due mainly to the appreciation of the currency, has shown a decline since 1991, similar to that of the early 1980s, despite the fact that the real exchange rate has been roughly constant. The balance on investment income, which rose until 1979, has trended down throughout the 1980s and 1990s, going negative for the first time in 1997. With the exception of 1991, when there was a $43 billion dollar transfer to the United States in payment for the Gulf War, unilateral transfers have also been a drag on the current account. Were it not for the tremendous growth in net exports of services since 1985, the downward trend in the current account balance would have been far greater.

The important question is what trends we should anticipate for these components of the current account in the future. Specifically, can we expect net service exports and capital income flows to continue to offset a good deal of the declining trade balance in goods? Further, what should we expect in terms of goods exports for the future?

Turning first to the trade in goods, it is important to understand that most of trade in goods is manufactured goods. In 1997, 85 percent of exports

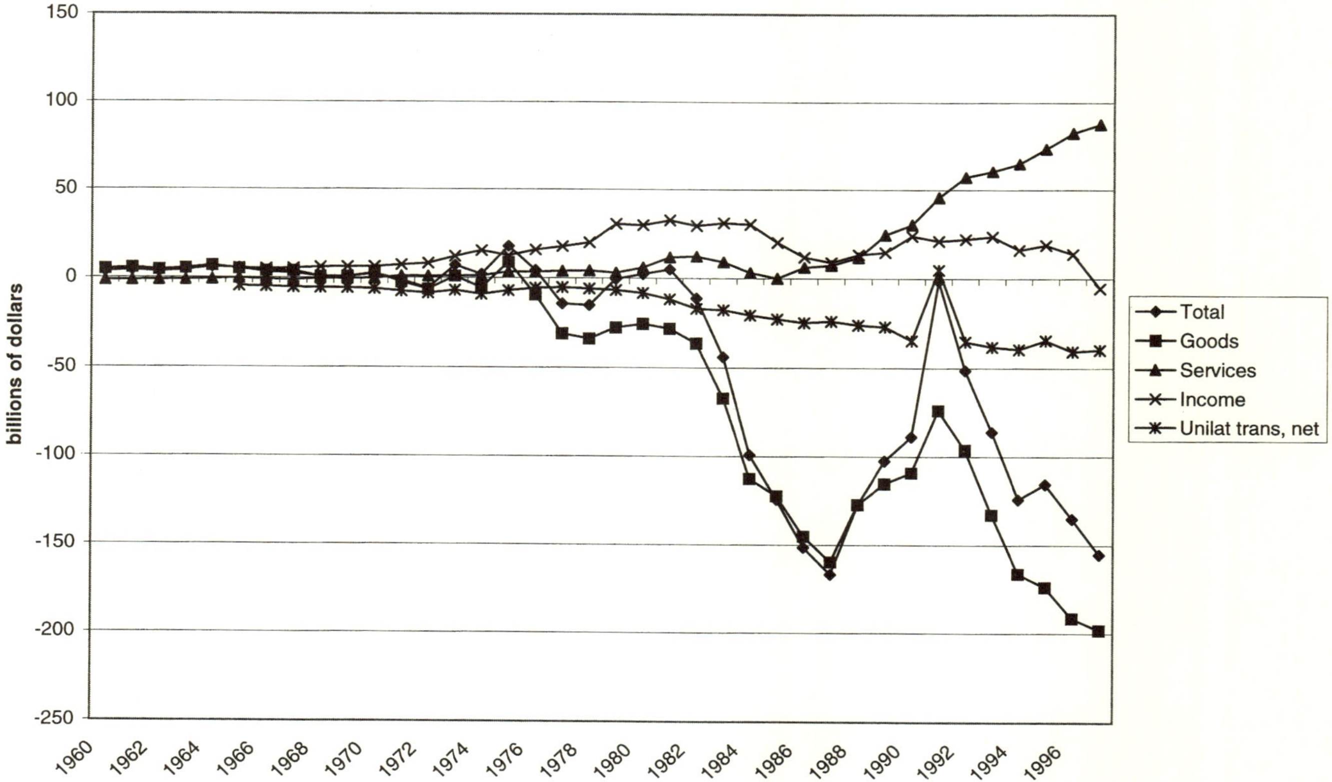

Fig. 9.2. Current account balance, 1960–97

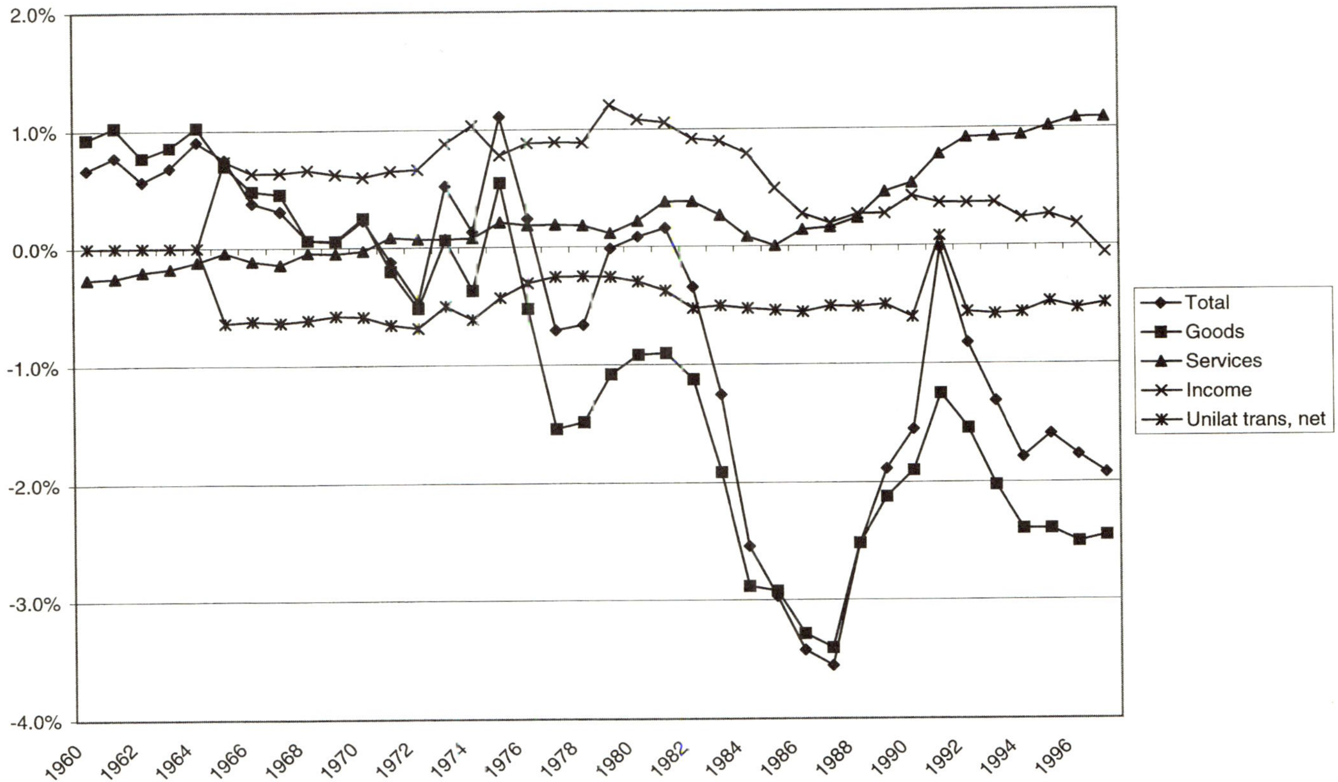

Fig. 9.3. Current account balance, as percentage of GDP

and 84 percent of imports were manufactured goods, up from 71 and 60 percent in 1978. To a considerable extent the rising share of manufactures probably reflects a decline in prices of primary goods relative to manufactured goods. Nonetheless, the trade balance in mineral fuel and agriculture products has been virtually unchanged since 1980.[9] Thus, as figure 9.2 shows, fluctuations in the goods trade balance since 1980 have been due entirely to fluctuations in the manufacturing trade balance.

Of course, much of the fluctuation in the trade balance between 1980 and 1990 was due to relative price fluctuations. Since 1990, however, relative prices or real effective exchange rates have been constant, and still there has been a significant decline in the manufacturing trade balance.[10] Thus the decline in the trade balance must be largely a function of the effect of relative rates of growth of U.S. and foreign demand, given the income elasticities of demand for U.S. imports and exports. Since U.S. and foreign demand have been growing at about the same rate since 1990, differences in the rates of growth of imports and exports can be a crude approximation of trends in nonprice competitiveness. It seems that manufactured imports and exports have grown (in nominal terms) at about the same rate since 1990. However, if we disaggregate manufactured trade by end-use categories, a disturbing trend emerges (table 9.2). First, consumer goods imports and exports have been growing at about the same rate over the period. While this suggests no negative trend in relative competitiveness, neither does it suggest an improvement. Since consumer imports already exceed exports by $115 billion, equal to 1.6 percent of GDP, we should expect to see a continued deterioration in the consumer goods trade deficit unless either U.S. relative prices fall or there is a marked improvement in world growth relative to the United States.

Automotive products exports, on the other hand, have grown significantly faster than automotive imports since 1990, probably reflecting the slowdown of Japanese parts imports and the effect of NAFTA on the export of fully assembled vehicles from the United States to Canada. Since, like consumer goods, automotive imports still exceed exports by about $66 billion, equivalent to 1 percent of GDP, we should not expect to see the automotive trade deficit shrink. However, its rate of growth will continue to slow.

What is especially disturbing, however, is the trend in capital goods trade. Capital goods imports have continued to grow at a faster rate than exports in the seven years since 1990, just as they did between 1980 and 1990. Since 1978, the capital goods share of manufacturing imports has risen from 19 percent to 35 percent, while their share of manufacturing exports has risen by only 3 points from 46 to 49 percent. Clarida and Hickok (1993) argued that such a trend in the relative competitiveness of capital goods could be expected to cause unfavorable trends in the relative income elasticities of

TABLE 9.2. Nominal Growth of Manufactured Imports and Exports, Disaggregated by End-Use Category, 1990–97 (index, 1990 = 100)

	1980	1990	1997
U.S. demand	76	100	113
Foreign demand	49	100	113
Man imports (M)	34	100	176
Man exports (X)	51	100	175
Consumer goods M	32	100	181
Consumer goods X	40	100	179
Capital goods M	*26*	*100*	*218*
Capital goods X	*49*	*100*	*192*
Auto products M	32	100	161
Auto products X	42	100	200

Source: Author's calculations from U.S. Department of Commerce, ITA 1998, table 4.

imports and exports. Demand for capital goods is generally considered to be more income-elastic than that of other manufactured goods. As the share of this sector in US exports stagnates, while its share in imports is rising (trends that are not driven by differences in relative rates of growth of demand), the weight of the sector in the calculation of elasticity will fall for exports and rise for imports. Recently, Hooper, Johnson and Marquez (1998) have provided some evidence that the income elasticity of demand for exports has fallen since 1985, though their study also concludes that the income elasticity of demand for imports has been stable since 1980.[11]

Thus, given the gap between the income elasticities of demand for imports and exports, and given that world demand is not now growing faster than that of the United States and that the relative prices of U.S. exports are not falling, further deterioration in the manufacturing trade deficit seems inevitable. Whether this can be compensated by trends in net capital income and net service exports is an obvious question, especially in light of the performance of these two components in the last seven years.

Net service exports have grown phenomenally since 1987 when the dollar returned to the real value of 1980 (fig. 9.2). Whereas the private services balance was $12 billion in 1980 and again in 1987, it had risen to almost $90 billion by 1997, substantially offsetting the nearly $200 billion deficit in goods. It has been suggested that the service surplus will continue to rise, alleviating the balance of payments constraint. However, if we disaggregate trends in the service sector, the prospects seem less encouraging. Whereas, in aggregate, service exports were growing considerably faster than imports between 1985 and 1991, when the United States was growing faster than its trading partners, since 1991, as rates of growth of aggregate demand have converged, so too have the rates of growth of service imports and exports. Of course, since service exports exceed imports by a considerable amount, the service trade surplus has continued to grow.

Disaggregation, however, reveals several discouraging trends. First, note that three categories—travel, royalties and license fees, and other private services—account for the entire surplus and about 70 percent of total service trade (fig. 9.4). Travel, as is clear from its dismal performance between 1981 and 1990, is extremely responsive to trends in the exchange rate. Thus, as the U.S. dollar has begun to appreciate against other currencies since 1995, the travel balance has not grown. The further appreciation anticipated with the collapse of emerging markets can be expected to depress the travel balance again. Royalties and license fees also seem to demonstrate sensitivity to trends in the value of the dollar. The really strong performer, of course, is "other private services," which includes the affiliated services of U.S. firms and their affiliates operating abroad, as well as unaffiliated financial services, telecommunications, and business, professional, and technical services. True, business, professional, and technical services, which were almost 30 percent of "other private services," almost doubled their nominal exports between 1991 and 1997, but imports grew even faster. With the exception of travel, in no category of service trade have exports been growing faster than imports since 1991. As will be shown in the next section, reasonable assumptions about the rate of growth of net service exports suggest that in 10 years net service exports will be about $200 billion, offsetting no more than about 25 percent of the projected merchandise deficit.

Finally, let us consider trends in the capital income flows portion of the current account. As figure 9.3 shows, at least from 1965 (when the data series begins) until 1984, net capital income flows averaged between 0.75 and 1 percent of GDP, more than offsetting the negative effect of the unilateral transfers, which were the only consistently negative component of the current account. Since its peak in 1978, however, the income balance as a percentage of GDP has been declining steadily. For the calendar year 1997, the net income flow on foreign assets and liabilities went definitively negative for the first time in history.

What was surprising was the fact that this did not occur sooner, given that U.S. net foreign assets had been negative since 1988 (fig. 9.5). But as Godley and Milberg (1994) had showed for the period before 1993, net investment income flows remained positive because the rate of return on U.S. investments abroad exceeded that of foreign investments in the United States. At the time, Godley and Milberg argued that it was reasonable to expect that if the United States sustained a trade deficit equal to about 1.7 percent of GDP, the rate of growth of net foreign debt would be large enough to generate

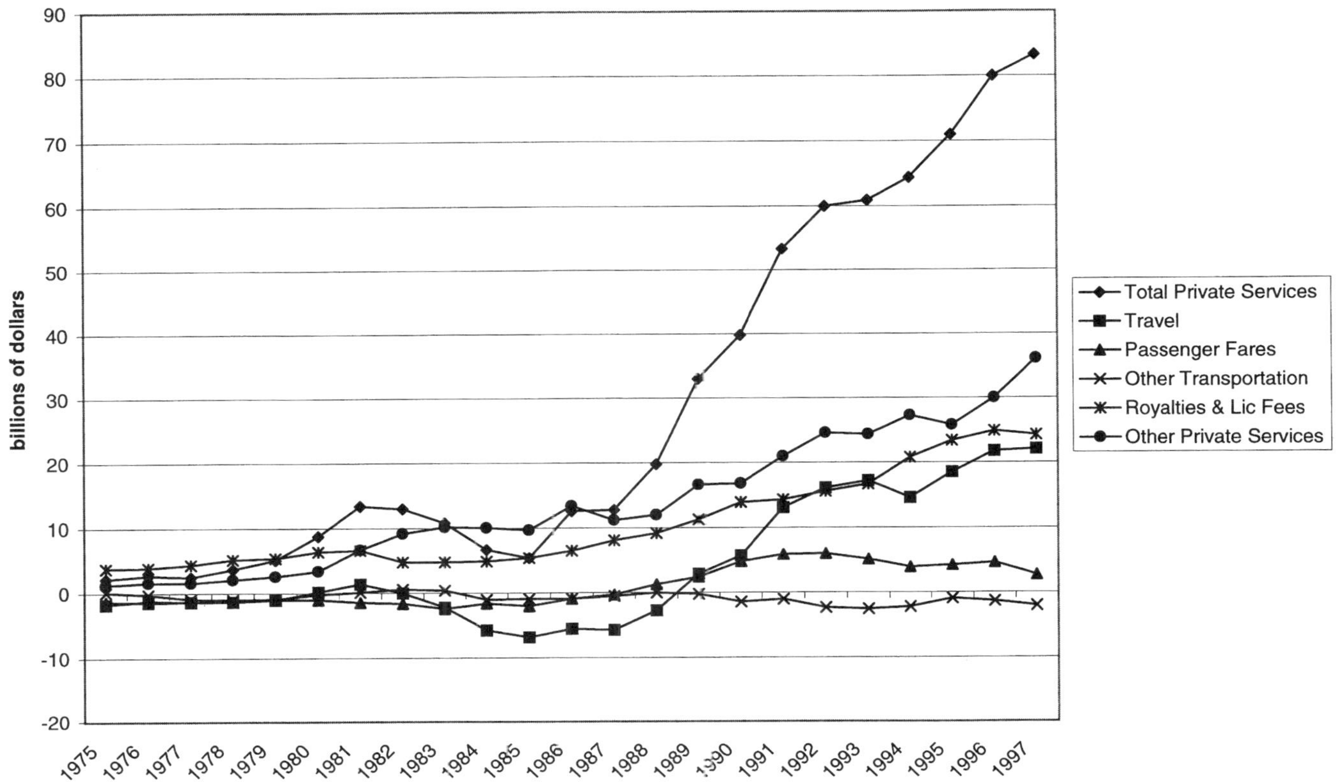

Fig. 9.4. Private services trade balance, 1975–97

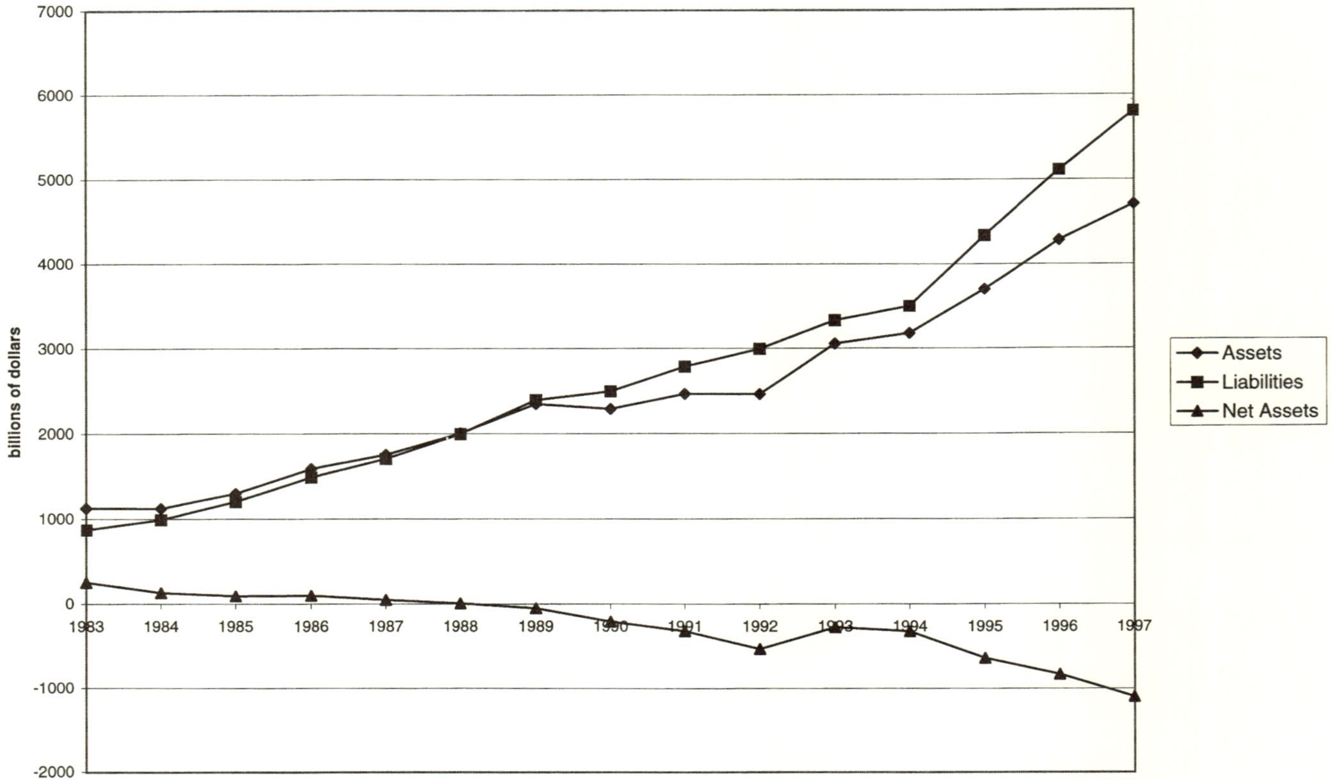

Fig. 9.5. Foreign assets and liabilities

negative net capital income flows that would within a few years be equal to 1 percent of GDP.

While Godley and Milberg's predictions did not bear out immediately, further analysis of trends in the composition and growth of foreign debt through 1997 confirms the general if not precise conclusions of their work. What it reveals is that given the rates of growth of foreign assets and liabilities that have prevailed since 1994, it does indeed seem likely that within five years U.S. net foreign liabilities could reach almost $7 trillion (table 9.3). Under reasonable assumptions about the rate of growth of GDP, the net liabilities would amount to about 70 percent of GDP. The net outflow of income on that volume of foreign liabilities could be as much as half a trillion dollars, or 6 percent of GDP.[12]

Table 9.3 illustrates how the above scenario might come to be. The first three columns show the value of U.S. assets abroad, and foreign assets held in the United States, between 1995 and 1997. It reveals that by 1997, U.S. net liabilities exceeded $1 trillion (Bach 1998). Also shown is the income earned by each asset category and the rate of return for each asset category in each of these three years. The overall rate of return earned by U.S. holders of foreign assets has exceeded the rate of return on U.S. assets held by foreigners for all three years. This is mainly because the rate of return on outward foreign direct investment, which represented about 35 percent of U.S. investors' foreign portfolio, was considerably higher than the rate of return earned by foreigners making direct investments in the United States. Column 4, however, shows that the rate of growth of FDI into the United States has exceeded the rate of growth of FDI abroad by U.S. investors for the last three years. If the rate of growth of each of these asset categories continues at the pace set in the last three years, and if the rate of return earned in each of these categories is sustained, then the scenario laid out above and illustrated in this table is a reasonably accurate prediction of future net capital income flows.[13]

Implications for External Balance and Growth

A simple simulation gives us some sense of the extent to which net service exports will offset the deterioration in the manufacturing trade balance (table 9.4). If we assume imports and exports for each of the categories of service trade continue to grow for the next 10 years at the same rate they have grown since 1991, then within 10 years, the service surplus will rise from $87 billion in 1997 to about $200 billion. However, if the same assumption is applied to goods components of the current account, within 10 years the merchandise trade deficit would be $777 billion. Thus, at best, the service trade will offset

TABLE 9.3. Trends in Net Foreign Assets and Capital Income Flows, 1995–97 and 2002 (in billions of dollars)

	1995	1996	1997	Growth Rate, 1995–97	2002 at 1995–97 Growth Rates
U.S. assets abroad	3,754	4,347	5,007	15.5%	10,663
U.S. government	257	242	216	–8.3%	140
FDI	1	1,517	1,794	17.1%	3,956
Other private	2,190	2,588	2,997	17.0%	6,566
Foreign assets in U.S.	4,291	5,091	6,330	21.4%	17,142
For official assets	672	801	834	11.4%	1,432
FDI	1,006	1,224	1,620	26.9%	5,341
Other private	2,614	3,066	3,875	21.8%	10,369
U.S. income on foreign assets	204	213	242		552
U.S. government receipts	5	5	3		2
FDI receipts	93	100	109		267
Other private	106	108	128		282
For income on U.S. assets	185	198	247		1,136
U.S. government payments	57	67	87		80
FDI payments	30	33	45		251
Other private payments	97	97	113		804
Net U.S. income	19	14	(5)		(585)
Net government receipts	(53)	(63)	(84)		(78)
Net income on FDI	63	66	64		16
Net income on other securities	9	11	15		(522)
U.S. ROR on foreign assets	5.4%	4.9%	4.8%		5.2%
U.S. government receipts	1.8%	1.9%	1.6%		1.6%
FDI receipts	9.3%	8.2%	6.8%		6.8%
Other private receipts	4.8%	4.2%	4.3%		4.3%
For ROR on U.S. assets	4.3%	3.9%	3.9%		6.6%
Payments to foreign gov'ts	5.1%	8.4%	5.6%		5.6%
FDI payments	3.7%	2.7%	4.7%		4.7%
Other private payments	8.1%	3.2%	7.8%		7.8%

Source: Compiled by author from U.S. Department of Commerce, Bureau of Economic Analysis 1998, table 3, p. 34; table 1, p. 68.

one-quarter of the goods deficit. More disturbing, however, are the implications of extending the analysis in table 9.3 for another 5 years. Within 10 years, net foreign liabilities would be $24 trillion dollars (in nominal terms), which would be about twice the value of GDP. The net negative income flow would be almost $2 trillion, which would be about 15 percent of GDP. The overall current account deficit would have reached $3 trillion or 25 percent of GDP (nominal). The service component of the current account, dynamic as it is, would be absolutely dwarfed by trends of this magnitude.

As a mild check to these absolutely astounding conclusions, I have also calculated the effect of an uncompetitive manufacturing sector on the current account, using estimates of income elasticities and assumptions about relative

TABLE 9.4. Projected Trade Deficit, Given Reasonable Assumptions about Growth of Goods, Services, and Income (in billions of dollars)

	Exports				Imports				Current Account Balance				
Year	Total	Goods	Services	Income	Total	Goods	Services	Income	Total	Goods	Services	Income	Net Transfers
1997	1,179	679	258	242	1,294.9	877	170	247	−155.2	−198	88	−5	−40
Growth rate	8.5%	8.5%	7.8%	—	10.0%	10.2%	6.3%	—	—	—	—	—	—
2007	2,673	1,532	549	1,209	3,448.3	2,310	313	3,120	−2,719	−778	236	−1,911	−82

Source: Author's calculations from U.S. Department of Commerce, ITA 1998.

TABLE 9.5. Current Account Implications of Relative Growth Rates, Assuming No Currency Depreciation (in percentages)

U.S. Growth	ROW Growth	CA/GDP_{t+10}
1.3	3.0	−2.2
2.5	3.0	−5.8
3.0	3.0	−7.3
4.0	3.0	−10
3.0	4.0	−6
4.0	4.0	−9
2.5	5.0	−3.4

Source: Author's calculations using income elasticities from Hooper, Johnson, and Marquez 1998.

rates of growth (table 9.5). The most recent estimates of income elasticities from Hooper, Johnson, and Marquez (1998) find that the elasticity for imports has been 1.8 over the period of analysis, while for exports it has been 0.8. Setting aside all the problems involved in assuming that elasticities are constant over long periods of time, we use these elasticities to estimate the effect of relative rates of growth of U.S. and foreign demand. In all cases, I have assumed, consistent with today's reality, that there would be no real exchange rate depreciation for the United States. The first four scenarios assume that demand for the rest of the world grows at 3 percent a year, as it has approximately since 1990. If the current account deficit is to remain at its current sustainable rate of 2.2 percent of GDP, either U.S. growth must average no more than 1.3 percent or there must be a steady depreciation of the real exchange rate. If the United States continues to grow at its current rate of 2.5 percent, then the current account, based on these income elasticities, would within 10 years rise to 5.8 percent of GDP. A slightly faster rate of growth of 3 percent for the United States will raise the deficit to 7.3 percent of GDP in 10 years. However, if the United States wants to grow at rates comparable to the 1960s, it can expect a 10 percent current account deficit within 10 years. The last two scenarios indicate the effect of faster world growth on the U.S. options for growth. Four percent rates of growth of demand, both for the United States and the rest of the world, would generate a current account deficit of about 9 percent of GDP within 10 years. However, it is important to keep in mind that Hooper, Johnson, and Marquez's elasticity estimates—like everyone else's—are measuring only the effect of relative income growth on the demand for tradable goods and services. No one is including income flows in their estimates.

Conclusion

As was argued in the first chapter of this book, unless the United States acts to improve the relative competitiveness of its export sector, it faces an exceedingly unattractive menu of policy options—some combination of currency depreciation, high interest rates, and slow growth—to control what may be a rapidly growing current account deficit.

Several possible scenarios for U.S. growth prospects in the next five to ten years are suggested by the analysis in this chapter. In the worst case scenario, the effect of the Asian financial crisis and a collapse in Brazil will be to put U.S. banks in an extremely vulnerable position. Expectations of a U.S. banking crisis will depress consumption and investment, sending the United States into a recession that will likely be followed by the rest of the world. Slow U.S. growth under those conditions will do nothing to improve the current account position as demand in the rest of the world will also be depressed.

A preferred scenario would see a fall in consumption, perhaps due to a change in tax policies, which would permit an increase in investment without increasing the current account deficit. Competitiveness would improve at the expense of Japan and Europe, and the current account deficit would begin to fall. However, as we argued in the introduction, given the present state of the world economy, were the United States to increase the competitiveness of its exports, without increasing its rate of growth, this would have the equivalent impact on world growth of imposing import restrictions. The rate of growth of our major trading partners, including Japan and other Asian countries, would be slowed as a consequence. Therefore, in the interests of long-run vitality, both of the U.S. and the world economies, it is necessary for the United States to increase both the competitiveness of its exports and its rate of growth simultaneously. For the United States to grow faster under this scenario, without worsening the current account deficit, would require substantial improvements in export competitiveness.

The best scenario would see faster growth in the EU and Japan, as well, as they resolve their current problems, and an increase in U.S. investment financed out of domestic consumption. Recovery in the rest of the world would both raise demand for the U.S. exports and slow the flow of portfolio capital into the United States as financial markets in Europe and Japan would become relatively more attractive to foreign investors. Net income flows out of the United States would slow, further strengthening the current account balance. If at the same time the United States undertook industrial policies targeted to industries that are strategic to the current account balance, such as electronics and auto, the structural imbalance in the external sector would be remedied that much faster.[14]

Thus under the best scenario, the United States would increase the propensity to export and reduce the propensity to import at a given rate of growth while simultaneously increasing its rate of growth. In such a scenario, U.S. action, especially if complemented by higher growth in Europe and Japan, would stimulate the rate of growth of the rest of the world because the volume of its imports would actually rise, even while its propensity to import fell.

NOTES

I wish to thank Ajit Singh, Lisa De Propris, and Ed McKenna for helpful comments and Irina Telyukova for her able research assistance.

1. As will be shown in the next section, the structural deterioration in the current account balance is a change from the 1950s and 1960s. Also to be shown in the next section is that the maximum rate at which the economy can grow, consistent with external balance, has been trending downward over time.

2. This is despite the fact that for generations, economists have estimated income elasticities as though they should appropriately be treated as constants over long periods of time.

3. It has been suggested, as well, that income elasticities pick up other factors that are not being measured in the price terms. For example, Helkie and Hooper (1988) found that if they included some measure of relative growth of productive capacity in the export and import equations, then the differential between the income elasticities all but disappeared. This, they argued, was a measure of the relative rates of growth of supply, which should have been reflected in a change in relative prices, but which for some reason—Cline (1989) argues mismeasurement—was not being captured in the price terms. It seems just as plausible, however, that relative rates of growth of productive capacity could reflect trends in relative nonprice competitiveness and demand for a country's products.

4. Relative export prices should capture the effects of all input prices, the exchange rate, and productivity growth differentials, although, to the extent that markets are not price competitive, relative export prices will also capture differences in profit margins or markup practices.

5. The table treats 1960 through 1969 as a single period, because it represents a full peak-to-peak business cycle. Trends in unit labor costs, however, are more appropriately viewed for the two subperiods, 1960 through 1965 and 1965 through 1969. During the first period, relative unit labor costs fell 5.3 percent a year as inflation rates abroad exceeded those in the United States. That was followed by a large increase in U.S. relative unit labor costs at an annual rate of 3.1 percent from 1965 to 1969 as U.S. inflation picked up and productivity growth slowed significantly (Hooper 1992, fig. 11.3). The current account responded accordingly, rising in the first half of the 1960s and falling after 1965.

6. The data used here for nominal effective exchange rates and relative unit labor costs for the period up to 1979 come from Hooper 1992. Measures of nominal and real and effective exchange rates (relative unit labor costs is the most commonly used REER) vary tremendously, depending on the price series being measured and the weights used. As an example of the differences weighting schemes can make, the OECD measures of nominal effective exchange rate trends and relative unit labor costs for U.S. manufacturing between 1969 and 1973 show an average decline of 5.1 for the former and 7.5 for the latter. Given the measurement problem, what is relevant for the discussion here is the general direction of the trend rather than precise magnitudes.

7. Hooper and Marquez (1995) show that during the 1980s, the appreciation of the dollar was fully passed through in lower import prices, but after 1985, the depreciation was not. Hence import prices remained below the 1980 relative level and have been falling since. The appreciation of the dollar was almost fully passed through in higher export prices as was the depreciation in lower export prices. Japanese pass-through is famously incomplete, weakening the link between exchange rate adjustment and changes in the real Japanese trade balance.

8. Since 1995, by some measures—unit labor costs and relative CPIs—real exchange rates have been trending up. However, by the measure of relative export unit costs, real exchange rates have been declining. Thus while the trend in exchange rates is somewhat ambiguous, most economists expect the value of the dollar in real and nominal terms to appreciate in the wake of the Asian financial crisis.

9. Given their relative insignificance as well as stability in the current account balance, agriculture and minerals are not likely to offset the manufacturing trade deficit unless there is a significant change in relative prices of commodities and manufactured goods.

10. Of course, while aggregate real effective exchange rates may have been constant, relative prices of different types of manufactured tradables may not have been constant. More careful analysis would require an analysis of the responsiveness of disaggregated trade flows to trends in disaggregated relative prices.

11. In a recent study they have estimated income and price elasticities for the G-7 countries, for 1955 through 1997 (depending on data availability), holding elasticities constant. However, for a shorter time period (1980 or 1985 to 1995), using a Kalman filter test, they tested for the stability of the income and price elasticities. For the U.S. they found that while the import income elasticities were constant at 1.8, export elasticities declined from about 1.4 to 0.8.

12. These calculations are done using nominal growth rates. Therefore, for the purposes of calculating the net income deficit as a percent of GDP, I have assumed that nominal income will grow at 4.5 percent a year, allowing a 2.5 percent real growth rate and 2 percent annual inflation.

13. It should be added that international capital markets are notoriously volatile and subject to dramatic reversals of flow over short time periods. Thus this scenario makes the heroic assumption that FDI in the United States as well as investment in U.S. securities and U.S. Treasury bonds will remain relatively more attractive than similar investments in other parts of the world. As long as U.S. real interest rates remain high relative to the rest of the world, and as long as much of the world's financial markets remain in turmoil, this is a reasonable assumption. However, the EMU could make investment in Europe sufficiently more attractive relative to the United States to slow the rate of investment and into and income flows out of the United States.

14. See Howes (1995) for a detailed discussion of which industries are strategic to achieving a sustainable external balance.

REFERENCES

Bach, Christopher L. 1998. U.S. International Transactions, Fourth Quarter and Year 1997. U.S Department of Commerce, *Survey of Current Business,* April: 51–97.

Blecker, Robert A. 1992. *Beyond the Twin Deficits: A Trade Strategy for the 1990s.* Armonk, N.Y.: M. E. Sharpe, Economic Policy Institute Series.

———. 1999. The Trade Deficit and US Competitiveness. In this volume.

Clarida, Richard H., and Susan Hickok. 1993. US Manufacturing and the Deindustrialisation Debate. *World Economy* 16, no. 2: 173–92.

Cline, William R. 1989. *United States External Adjustment and the World Economy.* Washington, D.C.: Institute for International Economics.

Dornbusch, Rudi. 1996. The Effectiveness of Exchange-Rate Changes. *Oxford Review of Economic Policy* 12, no. 3 (autumn): 26–38.

Godley, Wynne and William Milberg. 1994. U.S. Trade Deficits: The Recovery's Dark Side? *Challenge* 37, no. 6 (November/December): 40–47.

Helkie, William L., and Peter Hooper. 1988. An Empirical Analysis of the External Deficit In *External Deficits and the Dollar: The Pit and the Pendulum,* ed. Ralph C. Bryant et al. Washington, D.C.: The Brookings Institution.

Hooper, Peter. 1992. Macroeconomic Policies, Competitiveness and U.S. External Adjustment Chapter 11 in *International Productivity and Competitiveness,* ed. Bert G. Hickman. New York and Oxford: Oxford University Press.

Hooper, P., and J. Marquez. 1995. Exchange Rates, Prices and External Adjustment in the United States and Japan. In *Understanding Independence,* ed. P. Kenen, 107–68. Princeton, N.J.: Princeton University Press.

Hooper, Peter, Karen Johnson, and Jaime Marquez. 1998. Trade Elasticities for G-7 Countries. Board of Governors of the Federal Reserve System, International Finance Discussion Papers, no. 609, April.

Houthakker, H. and S. Magee. 1969. Income and Price Elasticities in World Trade. *Review of Economics and Statistics* 51:111–25.

Howes, Candace. 1995. Long-term Economic Strategy and Employment Growth in the U.S.: An Analysis of Clinton's Economic Policies. *Contributions to Political Economy* 14:1–31.

Howes, Candace, and Ajit Singh. 2000. In this volume.

Kaldor, Nicholas. 1978. The Effects of Devaluation on Trade in Manufactures. In N. Kaldor, *Further Essays on Applied Economics.* London: Duckworth.

———. 1981. The Role of Increasing Returns, Technical Progress and Cumulative Causation in the Theory of International Trade and Economic Growth. *Economie appliquée* 34, no. 4: 593–617.

Klitgaard, Thomas, and James Orr. 1998. Evaluating the Price Competitiveness of U.S. Exports. Federal Reserve Bank of New York. *Current Issues in Economics and Finance* 4, no. 2 (February): 1–6.

Krugman, Paul. 1989. Differences in Income Elasticities and Trends in Real Exchange Rates. *European Economic Review* 33:1031–54.

———. 1994. Competitiveness: A Dangerous Obsession. *Foreign Affairs* 73 (March/April): 28–44.

Lawrence, Robert Z. 1990. U.S. Current Account Adjustment: An Appraisal. *Brookings Papers on Economic Activity,* no. 2: 343–89.

Mann, Catherine L. 2000. Improving U.S. International Competitiveness: Macro Policy Management versus Managed Trade Policy. In this volume.

OECD. 1998. *OECD Economic Outlook,* no. 63 (June).

U.S. Department of Commerce, Bureau of Economic Analysis. 1998. *Survey of Current Business* (July).

U.S. Department of Commerce, ITA. 1998. *U.S. Foreign Trade Highlights* (June) Washington, D.C.: Government Printing Office.

Contributors

Robert A. Blecker is Professor of Economics at the American University, Washington, D.C. and a Research Associate of the Economic Policy Institute (EPI), Washington, D.C.

Candace Howes is the Barbara Hogate Ferrin '43 Associate Professor of Economics at Connecticut College, New London, Conn.

William Lazonick is University Professor of Economics at the University of Massachusetts, Lowell, and is on the faculty of the European Institute of Business Administration (INSEAD).

Daniel Luria is Director of the Performance Benchmarking Service, Industrial Technology Institute, University of Michigan, Ann Arbor.

Catherine L. Mann is a Senior Fellow with the Institute for International Economics in Washington, D.C. Her chapter was written while she was a Senior Economist at the Board of Governors of the Federal Reserve System.

Ann Markusen is Professor of Public Policy and Urban Planning at the University of Minnesota and Senior Fellow, Council on Foreign Relations, New York.

Mary O'Sullivan is on the faculty of the European Institute of Business Administration (INSEAD), Fontainebleau, France.

Ajit Singh is Professor of Economics at the Queen's College, University of Cambridge, U.K.

W. Edward Steinmueller is Professor of the Economics of Technical Change and Innovation, SPRU—Science and Technology Policy, University of Sussex, U.K.